Smart Operation for Power Distribution Systems

Daniel Pinheiro Bernardon
Vinícius Jacques Garcia
Editors

Smart Operation for Power Distribution Systems

Concepts and Applications

Springer

Editors
Daniel Pinheiro Bernardon
Technology Center
Federal University of Santa Maria
Santa Maria, Rio Grande do Sul, Brazil

Vinícius Jacques Garcia
Technology Center
Federal University of Santa Maria
Santa Maria, Rio Grande do Sul, Brazil

ISBN 978-3-030-06746-5 ISBN 978-3-319-93922-3 (eBook)
https://doi.org/10.1007/978-3-319-93922-3

This Springer imprint is published by the registered company Springer Nature Switzerland AG
The registered company address is: Gewerbestrasse 11, 6330 Cham, Switzerland

Preface

With the evident dependence of the electric power generation with fossil fuels, it is imperative to consider the inclusion of renewable energy sources in the composition of the energy matrix. This introduction highlights issues such as the need for greater control due to the more irregular generation profile and the need to incorporate technologies that promote the reduction of the distance between the point of generation and the point of consumption, with the desirable gains in efficiency and independence of generation.

The energy industry has contributed significantly to the technological advances mainly focused on the automation and evolution of information and communication technologies in order to make this control effective and efficient. Exactly from this promising context comes the basis of intelligent operation and mass-based computing controls that form the cornerstones of what is conventionally called smart grids.

The book addresses the subject of Operation of Electrical Distribution Systems, presenting contemporary concepts and real applications with focus on smart grids. The main techniques of active management of distribution systems operation are presented with examples of real applications, including state estimation, self-healing, volt-var control, protection systems, operations planning, commercial and emergency dispatch.

Chapter 1 presents an overview of the main concepts related to smart grid technologies and their interconnections with the operation of electrical distribution systems. The purpose is to offer an introductory basis for situating the reader in the context of the book. Topics include distributed generation, demand response, smart meters, electric vehicles, automation technologies and structures, and the role of distribution operation centers, together with some examples of automated processes in distribution networks.

Chapter 2 aims to describe state estimation issues, which have been widely discussed and applied in power systems since the early 1970s. Currently, in a context of greater dynamism in distribution systems, with the presence of distributed generation and other smart grid concepts, it became necessary to implement state estimators in the operation of these systems. Chapter 2 presents the classic weighted least squares method with the application in a simple numeric example. An introduction to other related issues such as observability and ill-conditioning is also provided, also including a review of other well-known techniques for state estimation.

Chapter 3 covers the characteristics and definitions related to the operation of self-healing systems in power distribution networks. The first part of the chapter presents an overview of the main concepts related to self-healing, as well as different network topologies that can arise depending on the signaled protection functions in the automated equipment or in the substation breaker. The second part addresses decentralized self-healing-based systems, which consider equipment performing actions based on their own information, independent of a central command or human interference. In the third part, centralized systems are described, which are dependent on the existing communication system between the field equipment and the SCADA system. Finally, some practical applications are presented, considering decentralized and centralized systems as examples.

Chapter 4 presents a review of the voltage and reactive power control (VVC) evolution and challenges applied in the smart grids. In the first part of the chapter, the VVC basic concepts are presented as well as the VVC objectives and challenges. In the second part are presented the main equipment and its operation characteristics. The third part comprises the VVC implementation and the main structures. Finally, an example is presented to demonstrate the application of VVC.

Chapter 5 provides an overview of overcurrent protection in power distribution systems. The main devices involved are presented, with their characteristics, functionalities, and the corresponding performances against faults, from protection theory. Next, methods of selectivity and coordination are discussed, which are used to relate their performance to network characteristics and the relationship between different types of equipment. In order to demonstrate the use of this theory, some practical examples are given. At the end, some issues about smart protection when considering distributed energy resources are provided.

Chapter 6 presents the challenges related to planning and managing the wide range of services inherent in an electrical distribution system. These services are performed by a set of teams, which are responsible for maintaining the electricity distribution grid fully functional and for serving the population. The chapter discusses the impact of the requirements imposed by the business context (types of services, customer criticality, regulatory agency penalties, etc.) in the process of planning the field activities of the teams. Important decisions, such as facility location, capacity planning, teams scheduling, and services assignment, are discussed. Understanding the decisions in the long-, medium-, and short-term horizons can help engineers and managers to improve the service quality while keeping the cost at acceptable levels.

Chapter 7 covers the dispatch problems in power distribution systems, presenting the decision-making problems and mathematical formulations related to the service crews and equipment operations. The first part of the chapter presents an overview of the dispatch problem, followed by the economic dispatch problem definition, which aims to minimize the cost of the generated power in order to balance supply-demand-storage. The latter discussion refers to scheduling, assignment, and routing in dispatch problems, with special attention to the vehicle routing problem. Herein is provided the offline, online, deterministic, and stochastic taxonomy and the structure of the dispatch problems. Overall in this chapter, correlated applications and their corresponding algorithms are discussed.

We wish a pleasant and fruitful reading, with the expectation that it will give rise to further theoretical and practical insights.

Santa Maria, Brazil

Daniel Pinheiro Bernardon
Vinícius Jacques Garcia

Acknowledgments

The authors would like to thank the technical and financial support of RGE Sul Power Utility (CPFL Energia) to the projects "Solução Inovadora para Gerenciamento Ativo de Sistemas de Distribuição" and "Planejamento Dinâmico de Operações" (P&D/ANEEL), Coordination for the Improvement of High Level Personnel (CAPES), National Center of Scientific and Technological Development (CNPq), and the National Institute of Science and Technology in Distributed Generation (INCT-GD) of Federal University of Santa Maria—UFSM, Brazil (CNPq process 465640/2014-1, CAPES process 23038.000776/2017-54, and FAPERGS 17/2551-0000517-1).

Contents

Chapter 1
Introduction to Smart Operation Centers

L. L. Pfitscher, A. R. Abaide, Daniel Pinheiro Bernardon, and Vinícius Jacques Garcia

1.1 Introduction

The first decades of this century has been characterized by significant increase in technological development, which has led to the modernization of many services and utilities. For example, one can cite the advances in telecommunications and in automotive transport, which have been substantially transforming these sectors. In addition, this period has also been marked by the population growth, the expansion of the industrial and commercial sectors, and a greater concern about environmental impacts produced by human action. This set of factors led to the need of rethinking one of the most important services for human development and quality of life: the electric energy.

The modernization of the Electric Power Systems has been occurring due to the need to meet the large-scale growth of electric energy consumption, taking into account new generation and load patterns, and making an intelligent use of the new technologies developed. This leads to a broader concept of operation of the power distribution networks which has been associated with the "smart grids."

Advances in the areas of automation and telecommunications are undoubtedly the driving force behind the establishment of the smart grids. They directly impact almost all of the services involved in the operation of the network and make the Distribution Operation Centers go through this process of transformation, leading to the concept of a smart network operation.

L. L. Pfitscher (✉)
Department of Energy and Sustainability, Federal University of Santa Catarina, Araranguá, Santa Catarina, Brazil
e-mail: luciano.pfitscher@ufsc.br

A. R. Abaide · D. P. Bernardon · V. J. Garcia
Technology Center, Federal University of Santa Maria, Santa Maria, Rio Grande do Sul, Brazil
e-mail: alzenira@ufsm.br; dpbernardon@ufsm.br; viniciusjg@ufsm.br

D. P. Bernardon, V. J. Garcia (eds.), *Smart Operation for Power Distribution Systems*, https://doi.org/10.1007/978-3-319-93922-3_1

The following sections present an overview of the most significant technologies and concepts related to the smart grids, the role of the automation, and new challenges to the distribution operation centers in this context.

1.2 Concepts of Smart Grids in Distribution Networks Introduction

In distribution networks, the term "smart grids" is used to characterize a set of techniques, methodologies, and procedures that aim to respond to the new demands and challenges of the Electric Power Systems, and that result in a significant transformation in their form of planning, operation, and control.

The smart grids are being driven by advances in automation and communication technologies, by the greater possibility of integration of the network equipment, and by the change in the way the services related to electric energy are treated. In this last aspect, issues such as billing, field team management, and reliability, among others, are included.

Various groups and work commissions on the subject were formed from the year 2000 on to establish integration patterns and discuss how conventional Power Systems, which were not thought as a "smart network," could gradually migrate to this new concept [1, 2]. Some factors accelerate the need for this transition, such as the growth of demand response devices, the greater integration of intermittent energy sources into the distribution and transmission systems, the impact of electric or hybrid vehicles on the networks, and the greater volume of information to be collected and processed from electronic power meters.

A scenario containing some typical elements of a smart grid is shown in Fig. 1.1. Three major areas are highlighted: the lower quadrant presents the services, technologies, and equipment of the power utility, including the physical energy network; the upper quadrant presents the consumers and related technologies; and the central area presents the technologies and equipment that may be common to the other two areas. The bidirectional arrows illustrate one of the main characteristics of the smart grids: the two-way communication and power flow. Some of the elements represented in Fig. 1.1 are discussed in the following sections.

1.2.1 Demand Response

The demand response refers to the capacity of managing and controlling loads of the power system. The primary objective of demand response is to reduce the peak power consumption by shutting down loads when the network is most demanded. Alternatively, consumption can be shifted to the periods of low demand, or may be

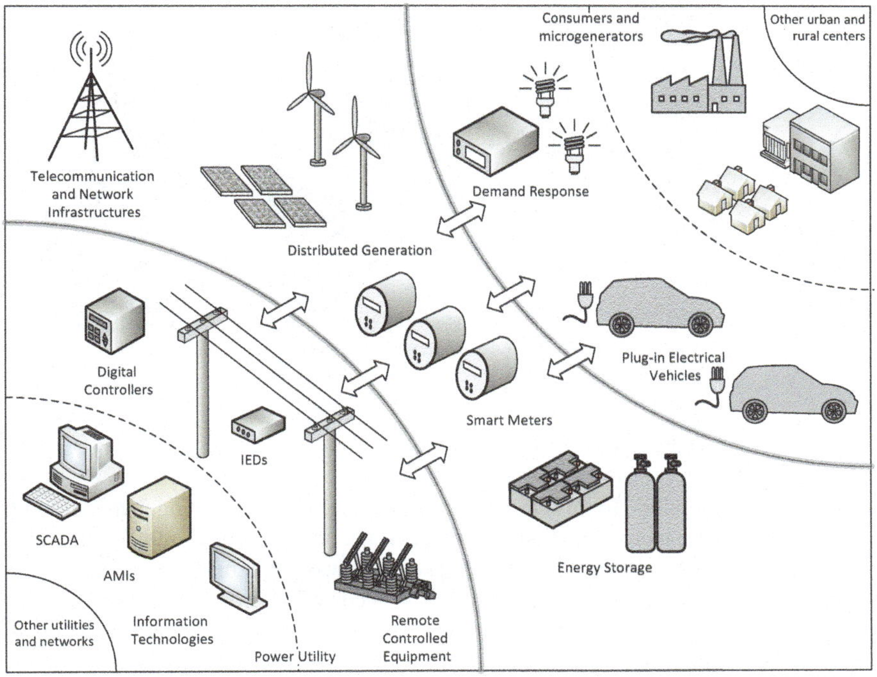

Fig. 1.1 Simplified representation of smart grids technologies

increased, to maintain generation stability or to take advantage of available resources from distributed generation [3].

In some cases, demand response may consider the possibility of transferring loads supplied by the power utility to a system of own generation of the consumer, to avoid cut of loads. However, this type of alternative can also be employed regardless of demand response, taking into account the type of contract of the consumer with the power utility.

The implementation of the demand response is done through electronic devices that communicate with the power utility. They send commands to disconnect the loads connected to grid, according to the request of the distribution operation center or depending on the variation of the energy tariff. This is the main difference between a demand response system and a conventional demand management system.

1.2.2 Advanced Metering Infrastructure (AMI)

The term "advanced metering infrastructure" (AMI) is actually used in different areas of supply and consumption of resources, such as electricity, gas, and water. The AMI involves all elements necessary for measurement and communication between

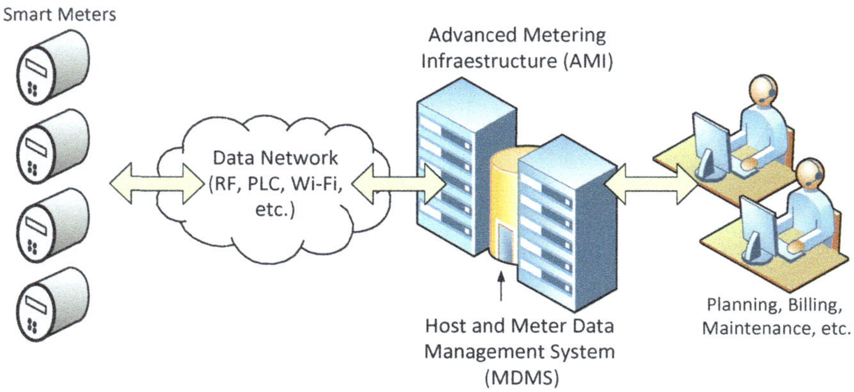

Fig. 1.2 Concept of advanced metering infrastructure

consumers and suppliers. In this case, the communication is bidirectional and allows a utility to send the consumer the financial value of the energy in real time, for example. The integration with demand response devices allows the management of loads according to price variation or according to the needs of the power utility [4].

AMIs are different from Automatic Meter Reading (AMR) systems that establish unidirectional communication and that are widely used for remote measurements, fault detection or diagnostic readings.

The concept of an Advanced Measurement Infrastructure is presented in Fig. 1.2. It includes electronic power meters, communication networks and a meter data management system (MDMS) which is responsible for storing and managing the large amount of data coming from the power meters, and establishing an interface between the collected data and other features of the network, such as billing systems and management of maintenance teams, for example. Some of the advantages of AMIs are the ability to track consumer billing in real time, the rapid detection of measurement failures and non-technical losses, and rapid response to energy reestablishment events. AMIs also allow the creation of more accurate consumer databases for profiling and demand estimation, and provide real-time measurements that aid in decision-making related to the system operation. Some technological challenges of deploying AMIs include the need for standardization of communication and interfaces between devices, and security issues that ensure that only authorized devices have access to the network information.

1.2.3 Smart Meters

The base of the Advanced Measurement Infrastructure is the Smart Meter, which can be considered as an evolution of the automated electronic power meter. The main feature of the smart meters is the bidirectional communication, which makes them able for real-time measurement and to receive commands from the power utility.

The communication standard varies depending on the application design of the meter. In some countries, for example, the power meters communicate by wire with a data concentrator and the data concentrator communicates with the distribution operation center by a wireless network. There are many power meters, however, that have the ability to communicate directly with the central by wireless.

Communication may be considered the greatest challenge of implementing smart meters in smart grids projects. There is a large variety of protocols and communication environment that can be used, and there is not a universal standardization of power meters. Some possible network arrangements consist on the use of cellular telephone networks, satellite communications, radiofrequency, Wi-Fi, and powerline communication (PLC), directly with the distribution operation center, making use of a data concentrator, or by cascading the communication through a mesh network. The main protocols used are defined by the ANSI C12.18 standard in the United States and the IEC 61107/62056 in Europe. Regardless of the type of communication, the discussion about the project to be adopted involves questions of cost, safety, and health, in the cases of equipment that emits radiofrequency radiation.

In terms of functionality, some common features in smart meters include the possibility of remotely connecting and switching off power, power outage alert, fraud alert (non-technical losses), real-time monitoring of the energy bill, and demand control. In some projects, smart meters also have the ability to communicate with consumer internal devices. For example, in a residence, the meter can receive or send information to and from appliances, air conditioners, etc. This concept, based on the home area network (HAN), allows the timely management of consumer loads.

1.2.4 Electric Vehicles

The development of the smart grids motivated the concern with changes in the consumer load profile, and among the most discussed points is the prospect of the increase of Plug-in Hybrid Electric Vehicle (PHEV) connected to the Power System [5].

Electric vehicles are characterized by the use of electric motors replacing fully or partially (in the case of hybrid vehicles) the traditional internal combustion engines for vehicle propulsion. Electric motors are usually powered by batteries. When the batteries are recharged through the power distribution grid, the vehicle is considered as "plug-in."

The growth of the PHEV fleet is justified by the vehicle's promising ability to improve efficiency, reduce fossil fuel dependence, and reduce environmental impact in terms of carbon emissions. However, its impact on the Electric Power Systems is significant, and several researches point out the need to plan the battery recharges to avoid the concentration of PHEVs overloading the system, and also solutions based on the use of the vehicle injecting energy into the network (V2G, Vehicle-to-Grid).

The relationship between PHEVs and smart grids lies in the fact that the grid must be prepared to absorb this new type of demand, which is not the case in conventional networks. The power imbalances (in the case of single-phase chargers), the increase of harmonic distortion in the network, and the increase of losses and voltage drops, especially in large feeders, can be cited as some aggravating factors of PHEVs in Power Systems. In addition, the capacity of network distribution transformers—in many cases already operating near their rated power—can be easily extrapolated if a set of PHEVs are simultaneously charging. The effects on the distribution network may also be reflected in the transmission system.

The smart grids can reduce the impact of PHEVs through supply and load management (demand response) and also by using differentiated pricing for supplying locations and times.

1.2.5 Distributed Generation

The concept of producing energy close to its place of consumption follows a trend of strong growth in the last two decades. This feature is associated with Distributed Generation (DG) where small generation plants replace or reinforce large power plants, often in cases when the cost of electricity transmission is high compared to the cost of the source implementation, or when the conventional power plant operates near its power limit. A general concept of distributed generation is that it refers to an energy source directly connected to the distribution network.

Distributed Generation usually employs wind and photovoltaic generators, due to the impossibility of building small conventional power plants such as hydroelectric or coal-fired power plants near the consumption centers. Gas and biomass thermal plants have also been used in GD, but on a smaller scale.

It is remarkable the increasing installation of photovoltaic panels in residences, because of their advantage of being modular, quiet, and more flexible to be installed on residential roofs—in the scale of up to 10 kW. Figure 1.3 illustrates a typical photovoltaic scheme connected to the grid, with the possibility of tariff compensation.

The growth of distributed generation has impacts on the conventional energy grid and also requires a network preparation for this new reality. The main technical aspects involved are related to power quality, stability, regulation capacity, and protection, due to the intermittent characteristics of the sources. Some of the features of the smart grids to support DG include advanced generation estimation, short- and long-term mechanisms, rapid power regulation, energy storage, and distributed resource management combined with demand response [6].

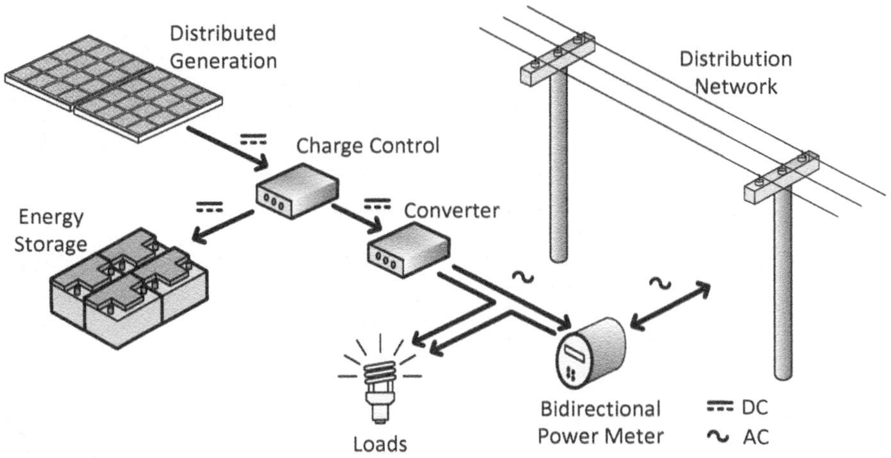

Fig. 1.3 Simplified performance model of a critical access module

1.3 The Role of Automation in Distribution Networks

Automatic systems usually involve monitoring and control actions constituting what is called a closed loop system, or a feedback system. In such systems, the information of the automated process is collected in field and passed to a device responsible for processing the information and sending commands to an actuator in order to maintain the operation of the process according to a predetermined setpoint. In the scope of the Distribution Systems, the automation can be applied to processes related to Operation, Maintenance, Planning, and Engineering, among others. Some of these automated processes are highlighted in Sect. 1.3.3. Figure 1.4 illustrates the scope of automation in distribution systems in the context of smart grids.

Among the advantages of automation in the distribution systems, the following stand out:

- Fast processing and exchange of information between the devices of the network, allowing real-time monitoring and control
- Reduction of costs and time with teams to carry out actions in field
- Possibility of creating more robust and reliable systems, less susceptible to human limitations
- Easier integration of network structures, such as distributed generation (DG), stability control, demand control, among themselves and with central monitoring and control systems

The benefits of automation, however, are often economically viable only if the automated system is part of a larger set, endowed with a certain degree of intelligence. This is the case, for example, of systems for automatic restoration of networks, in which the automatic actuation of remote switches must be integrated

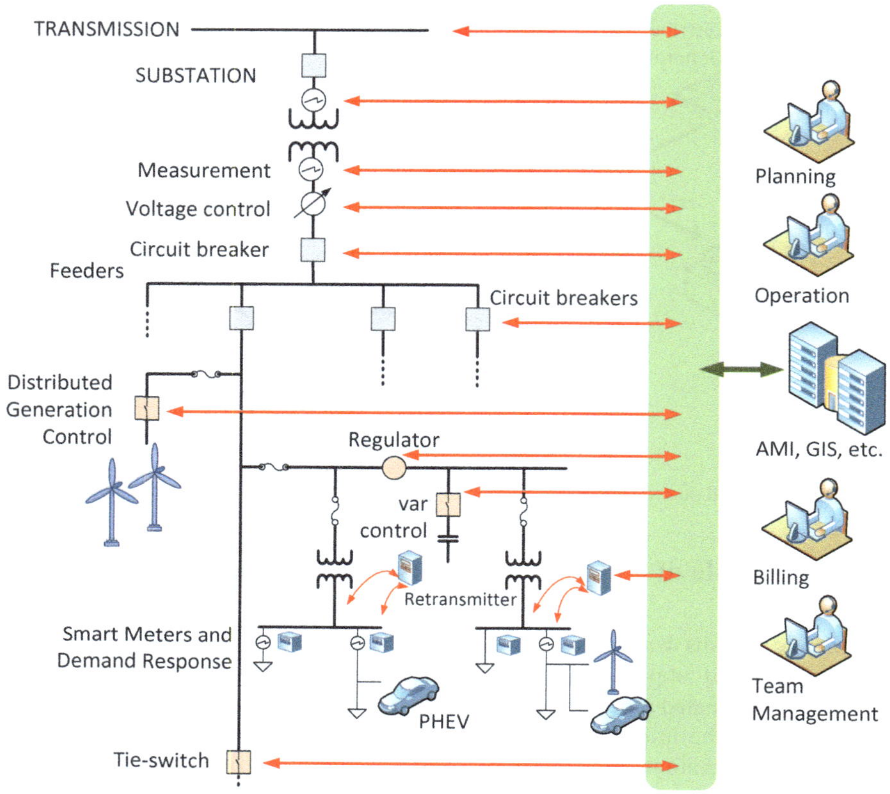

Fig. 1.4 Automation and communication in smart grid

with fault detection and isolation systems, guaranteeing the fast restoration of the higher possible number of consumers. In this case, the "intelligence" of the system can be represented by the response of computer programs that employ optimization and decision-making techniques.

It has grown significantly in recent years the employment of Artificial Intelligence, not only in Distribution Systems, but also in automated processes in general, whether in industry, commerce, or in personal electronic equipment.

Among the techniques of Artificial Intelligence, it is highlighted the use of Artificial Neural Networks, Genetic Algorithms, Evolutionary Programming, and Expert Systems. These techniques can be associated to other techniques, such as Heuristic and Fuzzy Systems, for a series of applications, such as: estimation of load demand or generation curves, estimation of the network state from real-time measurements, optimization of automatic equipment and DG sources, optimization of network reconfiguration and restoration, optimization of generation dispatch, recognition of power consumption patterns, and support in expansion planning. Some applications are discussed in the other chapters of this book.

1.3.1 Basic Structure of Automated Networks

The typical configuration of an automation system includes the following elements:

- Sensors and transducers: responsible for measuring the physical quantities of interest in the process and converting them into signals that can be processed. These are the main element of data acquisition systems such as those found in power meters.
- Electronic control devices: responsible for receiving information from sensors and transducers and other electronic devices, storing data, executing programming codes, receiving commands from other devices, and sending commands to other devices and actuators. Some examples of control devices include: programmable controllers, remote terminal units (RTUs), and intelligent electronic devices (IEDs).
- Actuators: responsible for receiving commands from control devices and turning them into physical action, such as turning on or off an equipment, and opening or closing a switch.
- Human-Machine Interface (HMI): responsible for providing communication between the user/operator and the control equipment. This communication can be for programming or changing local parameters of the equipment or the process, manual commands, monitoring and verification of the state of variables, among others.
- Supervisory systems: responsible for integrating functions of monitoring, alarm management, data storage and history, communication with devices, user interface, among others. These systems are commonly known by the acronym SCADA, and they are discussed in more detail in Sect. 1.3.2.

In addition, automated network systems also include two basic elements:

- Communication systems: include the physical medium of data communication, such as cables, radio, telephony, fiber optics, satellite link, and Wi-Fi, and network protocols such as TCP / IP, DNP, IEC 61850, and Ethernet, among others.
- Communication equipment: includes the devices responsible for receiving and transmitting information between the electronic equipment of the network and the SCADA System, when available. The configuration depends on the communication medium used and may include modems, data concentrators, encryption devices, and gateway converters.

Interoperability between communications standards and protocols plays a critical role in advancing smart grids. An important reference guide is presented by the National Institute of Standards and Technology (NIST) [7].

Figure 1.5 illustrates an example of architecture of an automated system.

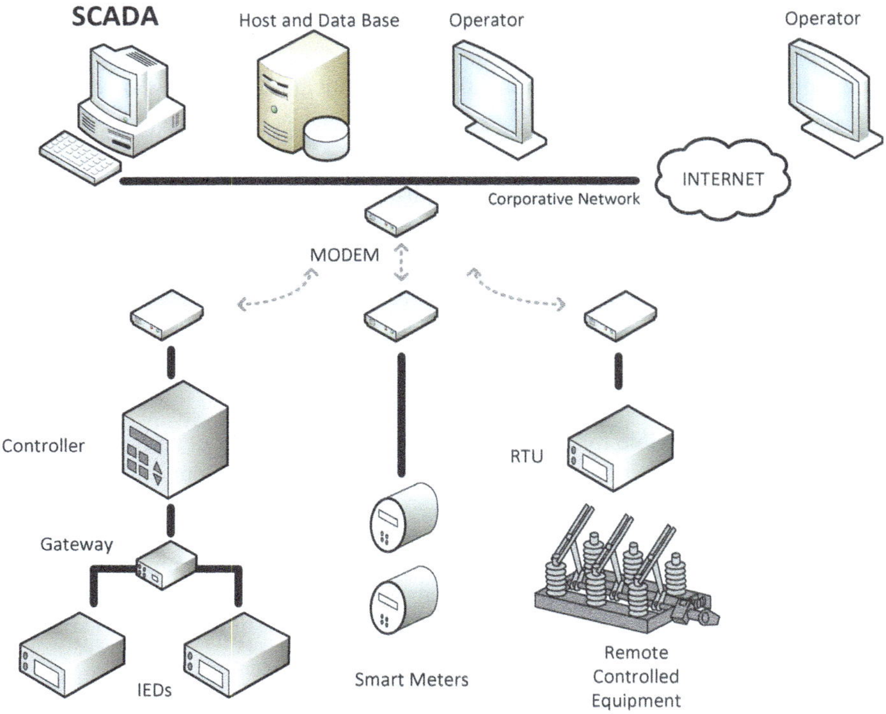

Fig. 1.5 Simplified architecture of an automated network

1.3.2 SCADA Systems

A SCADA (Supervisory Control and Data Acquisition) consists of a computer program that performs monitoring and control functions in an automation system. In general, this is a highly customized program, according to the characteristics of the plant or process under supervision. It represents the main interface between the user (supervisor or operator) and the processes. In Power Systems, the program typically includes specific and advanced application modules that integrate the Energy Management Systems (EMS) and Advanced Distribution Management Systems (ADMS).

Fig. 1.6 shows the context of the SCADA in a general network operation task.

Current SCADA systems predominantly use object-oriented programming platforms. From this platform, the user or programmer sets up a set of operational blocks that will constitute the application of the supervisor.

A typical SCADA application has the following basic elements:

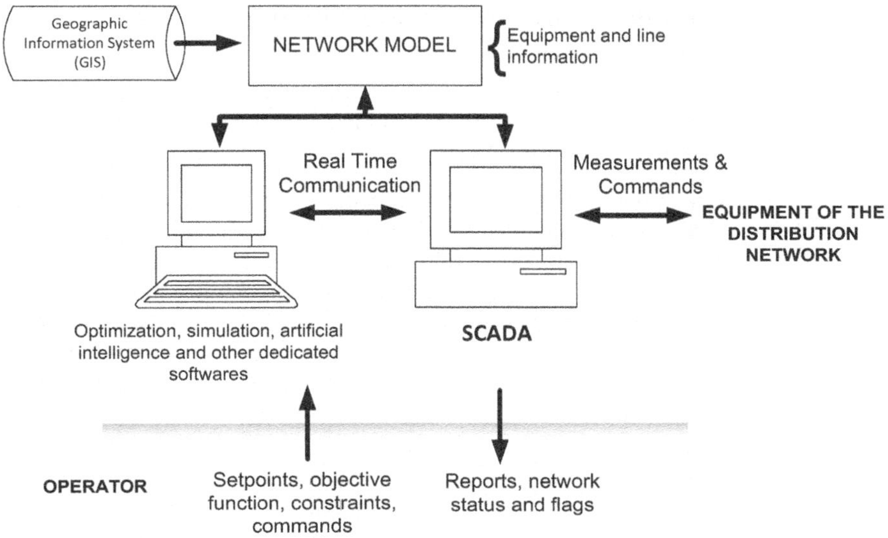

Fig. 1.6 Interface of SCADA with other power utility systems

- External interface: implemented through communication drivers, which connect the program to external hardware and software devices, such as modems, controllers, web servers, and other computer programs.
- Process screens: provide access to real-time information on plant variables and parameters, as well as interfaces to field devices, such as buttons and setpoints. The screens can be configured according to the user who is using the application.
- Constants and variables: all process parameters and screens can be associated with a SCADA element usually known as a "tag." Examples of tags are: values of physical process quantities, configuration parameters of the program communication system, screen configuration parameters, and screen objects (color, size, etc.), among others.
- Alarms: used to monitor events such as exceedance of limit values, change of state of digital variables, and abrupt variations of monitored quantities.
- Historical: store and provide information about plant records. They are usually linked to a database.

A SCADA/ EMS or SCADA/ ADMS system includes specific modules such as power flow calculation, representation of single-line diagrams, control of network devices (remote controlled switches, reclosers, capacitor banks, etc.), dispatch control, and operation.

1.3.3 Examples of Automated Processes in Smart Grids

The automation is absolutely necessary in almost all operation tasks of the smart grids. The following chapters of this book present in detail and discuss some of these operations, which are briefly introduced in this section:

- Self-restoration: in the event of a permanent fault in any part of the network, the smart grid must be able to quickly identify and isolate the region with the fault, and restore power to as many consumers as possible in the non-defective areas. This can be done by transferring loads between feeders or substations, aided by the connection of emergency generation systems. This process is related to a system known as FDIR (Fault Detection, Isolation and Restoration) or FLISR (Fault Location, Isolation and Service Restoration). Rapid power restoration has a significant impact on service quality indicators such as average duration and frequency of interruption (SAIDI and SAIFI).
- Automatic reconfiguration: during normal operating mode, smart grids must be able to automatically identify the topology of the network that presents the best performance according to optimization criteria such as loss minimization and improvement of quality indicators, considering the different scenarios of generation and consumption that occur over a period. After determining the best configuration, the reconfiguration should also be automatic, by the activation of remotely controlled devices (switches) in the network.

The combination of self-restoration and automatic reconfiguration leads to the concept known as self-healing, that is, the self-regeneration of the network to increase the reliability of energy supply.

The planning stage of the automation of a Distribution Network is important to define the strategic points of reconfiguration of this network, considering different communication technologies, such as cellular, radio, and fiber optic.

- Voltage and reactive (V/VAR) monitoring and control: this function is related to the control of capacitor banks, distributed generators, voltage regulators, and taps of network transformers, with the primary objective of establishing adequate levels of voltage and power factor and, as a result, improve power quality and reduce losses. The control is usually done in an integrated way, since the voltage regulation and the control of the reactive power are closely related.

To perform these operations, the network infrastructure should support: data management of power meters (MDMS), monitoring and control of network status (overload, reactive control, etc.), load and distributed generation management, charging control of plug-in electrical vehicle, among other features. Information systems should communicate among themselves, at different levels of application, such as power regulation strategies, service pricing, management of maintenance teams, database updating, and geographic information systems (GIS), among others [8].

1.4 The Distribution Operation Center

In energy distribution systems, the basic function of the Distribution Operation Center (DOC) of power utilities is to guarantee the supply of energy to all consumers connected to the system, within pre-established power quality standards. Typical DOC actions include: load forecasting, scheduling interventions in the physical network, load control in contingency or emergency situations, quality control of power supply, and operational coordination of systems.

Physically, the DOC structure consists of an advanced networked computing system in which operators and supervisors continuously monitor the state of the grid, make decisions about system interventions, and perform other coordination, command and execution of functions for the operation of the system.

An intervention comprises any action in the distribution system, such as network expansion, installation of equipment, execution of maintenance services, and testing the system itself.

The operations are normally regulated by internal standards of the power utilities and can be classified into three types:

- Pre-operation: includes previous studies about the interventions in the network, in order to minimize risks and increase reliability. At this stage, programs to support decision-making and management of work teams in the field are important.
- Real-time operation: includes energy monitoring systems (voltage levels, line and transformer loading, frequency, faults, etc.) and their controls.
- Post-operation: involves the creation of reports and database of occurrences in the network, which allow the elaboration of diagnoses and studies for network improvement.

In the following chapters of this book, the concept of smart distribution operation center is presented, and all operations mentioned are detailed, according to the specific application.

1.5 Conclusions

The features available with the advances in smart grids have brought advantages to the operation of distribution networks. Some of these advantages include the fast processing and exchange of information between the devices of the network, allowing real time monitoring and control; the reduction of costs and time with teams to carry out actions in field; the reduction of technical and non-technical losses; the possibility of creating more robust and reliable systems, less susceptible to human limitations; and easier integration of network structures, such as distributed generation sources, demand response, smart meters, PHEV chargers, and others. The benefits of technology, however, are often viable only if it employs an automated system endowed with a certain degree of intelligence, such as optimization software, for example. In the scope of the Distribution Networks, some resources of the smart grids stand out, either by the direct impact on the generation, consumption and quality indicators or by the impact in the network monitoring, operation and planning.

References

1. Brown RE (2008) Impact of smart grid on distribution system design In: Power and Energy Society General Meeting - IEEE Conversion and Delivery of Electrical Energy in the 21st Century. Pittsburg, pp 1–4
2. Ipakchi A, Albuyeh F (2009) Grid of the future. IEEE Power Energy Mag 7(2):52–62
3. Rahimi F, Ipakchi A (2010) Demand response as a market resource under the smart grid paradigm. IEEE Trans Smart Grid 1(1):82–88
4. Bouhafs F, Mackay M, Merabti M (2012) Links to the future: communication requirements and challenges in the smart grid. IEEE Power Energy Mag 10(1):27–28
5. Boulanger AG, Chu AC, Maxx S, Waltz DL (2011) Vehicle electrification: status and issues. Proc IEEE 99(6):116–1138
6. Atzeni I, Ordóñez LG, Scutari G, Palomar DP, Fonollosa JR (2013) Demand-side management via distributed energy generation and storage optimization. IEEE Trans Smart Grid 4(2):866–875
7. National Institute of Standards and Technology (2012) NIST framework and roadmap for smart grid interoperability standards, Release 2.0
8. Heidari S, Fotuhi-Firuzabad M, Lehtonen M (2017) Planning to equip the power distribution networks with automation system. IEEE Trans Power Systems 32(5):3451–3460

Chapter 2
State Estimation

R. G. Milbradt and L. N. Canha

2.1 Introduction

State estimation for power systems was first formulated as a weighted least-squares problem by Schweppe [1] in early 70s and has become an integral part of power system monitoring and operation. State estimation is a mathematical procedure to process the set of real-time measurements to come up with the best estimate of the current state of the system. By minimizing the sum of the squares of the differences between the estimated and the measured values of the system, a best estimate of the system is generated. Because state estimation for alternating current (AC) power systems is a nonlinear problem, traditionally state estimation has been solved using iterative methods such as the weighted least-squares [2]. The result of state estimation provides the real-time database for other applications, such as security assessment and control, and economic dispatch. An accurate state estimator is very important, and the result is the backbone of the grid planning and the power system operation. Large errors in the estimation may cause severe flaws in areas such as economic dispatch of power, transient and voltage stability, and the protection system of the grid. The increase of distributed generation results in distributed energy resources (DERs) and load dynamics in the distribution network [3]. Algorithms developed for state estimation in transmission systems need to be adapted to be suitable for distribution systems, that is, where DERs are installed and network observability is not achieved unless pseudo-measurements are used.

R. G. Milbradt (✉)
Polytechnic College, Federal University of Santa Maria, Santa Maria, Rio Grande do Sul, Brazil
e-mail: rafael.milbradt@ufsm.br

L. N. Canha
Technology Center, Federal University of Santa Maria, Santa Maria, Rio Grande do Sul, Brazil
e-mail: lucianecanha@ufsm.br

D. P. Bernardon, V. J. Garcia (eds.), *Smart Operation for Power Distribution Systems*, https://doi.org/10.1007/978-3-319-93922-3_2

In the next sessions the classical formulation of the state estimation problem is presented along with a practical example that allows readers to reproduce the formulation and replicate the technique for state estimation in real power systems. In addition, classic problems of the state estimation such as observability and ill-conditioning are discussed, which are problems that need attention when implementing a state estimator. In this chapter we also present a brief review of other state estimation techniques that have emerged from Schweppe's classical formulation.

2.2 Problem Formulation

The state estimators attempt to obtain the best estimate for the state of an electric power system, given the electric model of the network, as well as a set of related measures, each with a degree of accuracy, representing the reliability level of each of these measures. Typically, measures can be:

- Voltage magnitude (V)
- Real Power Flow (P_{flow}) and Reactive Power Flow (Q_{flow})
- Real Power Injection (P_{inj}) and Reactive Power Injection (Q_{inj})

And in some special cases:

- Current (I)
- Voltage phase angle (θ)

The state of a power system is typically represented by:

- Voltage magnitude in each bus/node (V)
- Voltage phase angle in bus/node (θ)

There are several types of state estimators for power systems, which will be listed in Sect. 2.5. In this section the mathematical formulation of the classic method in the literature, which was proposed by Schweppe [1] and known as Weighted Least Squares (WLS), will be presented. The WLS was the first that emerged in the early 1970s, and was adopted by several control centers worldwide. The classical formulation obtains the solution by Normal Equations via the Gaussian method. Further details on formulations and implementations, among other details, can be found in [4, 5].

2.2.1 *Weighted Least Squares (WLS) Method*

The WLS method consists of an optimization technique that tries to find the best state estimation of the electric network, based on the electric model of the network and in a set of measures associated to its weights. Thus, the estimate is obtained through:

$$\min j(x) = \sum_{i=1}^{m} W_i (z_i - h_i(x))^2 \tag{2.1}$$

where:
z_i: is the measurement i
W_i: is the weight of the measurement i
$h_i(x)$: is the function h that relates the state variable x and the measurement i
m: is the total of measurements

With the solution of this optimization problem we obtain the estimated state x which must satisfy the condition:

$$\frac{\delta J(x)}{\delta x} = 0 \Rightarrow H^T(x) W [z - h(x)] = 0 \tag{2.2}$$

where:

$$H(x) = \frac{\partial h(x)}{\partial x} \tag{2.3}$$

Therefore, $H(x)$ is the Jacobian matrix of $h(x)$ functions that relate the measurements to their respective states. The solution of the nonlinear Eq. (2.2) can be obtained through an iterative method where a linear equation is solved at each iteration to calculate the correction $\Delta x^{k+1} = x^k + \Delta x^k$.

$$\left[G\left(x^k\right)\right] \Delta x^{k+1} = H^T\left(x^k\right) W^{-1} \left[z - h\left(x^k\right)\right] \tag{2.4}$$

where G is called Gain matrix, and is defined by:

$$G(x) = H^T\left(x^k\right) W^{-1} H\left(x^k\right) \tag{2.5}$$

The matrix $G(x)$ is sparse, positive-definite, and symmetric, characteristics that are obtained when the system is completely observable (discussed in Sect. 2.4).

2.2.2 *Measurement Functions*—h(x)

The measurement functions matrix $h(x)$ is formed as follows:

$$h = \begin{bmatrix} V_{\mathrm{mag}} \\ P_{\mathrm{inj}} \\ Q_{\mathrm{inj}} \\ P_{\mathrm{flow}} \\ Q_{\mathrm{flow}} \\ I_{\mathrm{mag}} \end{bmatrix} \tag{2.6}$$

- Voltage magnitude at node j:

$$h(x) = V_j \tag{2.7}$$

- Real Power Injection at node i:

$$P_{i,\mathrm{inj}} = V_i \sum_{j=1}^{N} V_j \left(G_{ij} \cos\theta_{ij} + B_{ij} \sin\theta_{ij}\right) \tag{2.8}$$

- Reactive Power Injection at node i:

$$Q_{i,\mathrm{inj}} = V_i \sum_{j=1}^{N} V_j \left(G_{ij} \sin\theta_{ij} + B_{ij} \cos\theta_{ij}\right) \tag{2.9}$$

- Real Power flow from node i to node j:

$$P_{ij,\mathrm{flow}} = \begin{pmatrix} \alpha_{ij} V_i^2 \left(g_{si} + g_{ij}\right) - \\ V_i V_j \left(g_{ij} \cos\theta_{ij} + b_{ij} \sin\theta_{ij}\right) \end{pmatrix} \tag{2.10}$$

Where:
Reactive Power flow from node i to node j:

$$Q_{ij,\mathrm{flow}} = \begin{pmatrix} -\alpha_{ij} V_i^2 \left(b_{si} + b_{ij}\right) \\ -V_i V_j \left(g_{ij} \sin\theta_{ij} - b_{ij} \cos\theta_{ij}\right) \end{pmatrix} \tag{2.11}$$

- Current from node i to node j:

$$I_{ij} = \frac{\sqrt{P_{ij,\mathrm{flow}}^2 + Q_{ij,\mathrm{flow}}^2}}{V_i} \tag{2.12}$$

where:
$G_{ij} + jB_{ij}$: complex admittance at ij in the nodal admittance matrix
$g_{ij} + jb_{ij}$: series complex admittance of the line connecting node i to j

$g_{si} + jb_{si}$: shunt complex admittance of the line connected at node i
α_{ij}: transformer tap ratio, if exists, between nodes i and j

2.2.3 *Jacobian Matrix*—H(x)

The data structure of the Jacobian matrix adopts a strategy that seeks to facilitate its factorization and can be represented as:

$$H = \begin{bmatrix} \dfrac{\partial V_{\text{mag}}}{\partial \theta} & \dfrac{\partial V_{\text{mag}}}{\partial V} \\ \dfrac{\partial P_{\text{inj}}}{\partial \theta} & \dfrac{\partial P_{\text{inj}}}{\partial V} \\ \dfrac{\partial Q_{\text{inj}}}{\partial \theta} & \dfrac{\partial Q_{\text{inj}}}{\partial V} \\ \dfrac{\partial P_{\text{flow}}}{\partial \theta} & \dfrac{\partial P_{\text{flow}}}{\partial V} \\ \dfrac{\partial Q_{\text{flow}}}{\partial \theta} & \dfrac{\partial Q_{\text{flow}}}{\partial V} \\ \dfrac{\partial I_{\text{mag}}}{\partial \theta} & \dfrac{\partial I_{\text{mag}}}{\partial V} \end{bmatrix} \tag{2.13}$$

Each element of the Jacobian matrix is expressed by the differential equation of each measurement functions in relation to the state variables, angle and voltage magnitude. The matrix $H(x)$ is set up in way that each line represents a measurement and each column a state variable.

- Voltage magnitude:

$$\frac{\partial V_i}{\partial V_i} = 1, \frac{\partial V_i}{\partial V_j} = 0, \frac{\partial V_i}{\partial \theta_i} = 0, \frac{\partial V_i}{\partial \theta_j} = 0 \tag{2.14}$$

- Real power injection measurement P_{inj} for the k-th iteration:

$$\begin{aligned} \frac{\partial P_i^{k+1}}{\partial \theta_i^{k+1}} &= \sum_{j=1}^{N} \left(V_i^k V_j^k \left(-G_{ij} \sin\left(\theta_{ij}^k\right) + B_{ij} \cos\left(\theta_{ij}^k\right)\right) - V_i^{k^2} B_{ii}\right) \\ \frac{\partial P_i^{k+1}}{\partial \theta_j^{k+1}} &= V_i^k V_j^k \left(G_{ij} \sin\left(\theta_{ij}^k\right) - B_{ij} \cos\left(\theta_{ij}^k\right)\right) \\ \frac{\partial P_i^{k+1}}{\partial V_i^{k+1}} &= \sum_{j=1}^{N} \left(V_j^k \left(G_{ij} \cos\left(\theta_{ij}^k\right) + B_{ij} \sin\left(\theta_{ij}^k\right)\right) + V_i^k G_{ii}\right) \\ \frac{\partial P_i^{k+1}}{\partial V_j^{k+1}} &= V_i^k \left(G_{ij} \cos\left(\theta_{ij}^k\right) + B_{ij}^k \sin\left(\theta_{ij}^k\right)\right) \end{aligned} \tag{2.15}$$

- Reactive power injection measurement Q_{inj} for the k-th iteration:

$$\begin{aligned}
\frac{\partial Q_i^{k+1}}{\partial \theta_i^{k+1}} &= \sum_{j=1}^{N}\left(V_i^k V_j^k\left(G_{ij}\cos\left(\theta_{ij}^k\right) + B_{ij}\sin\left(\theta_{ij}^k\right)\right) - V_i^{k^2} B_{ii}\right) \\
\frac{\partial Q_i^{k+1}}{\partial \theta_j^{k+1}} &= V_i^k V_j^k\left(-G_{ij}\cos\left(\theta_{ij}^k\right) - B_{ij}\sin\left(\theta_{ij}^k\right)\right) \\
\frac{\partial Q_i^{k+1}}{\partial V_i^{k+1}} &= \sum_{j=1}^{N}\left(V_j^k\left(G_{ij}\sin\left(\theta_{ij}^k\right) - B_{ij}\cos\left(\theta_{ij}^k\right)\right) + V_i^k B_{ii}\right) \\
\frac{\partial Q_i^{k+1}}{\partial V_j^{k+1}} &= V_i^k\left(G_{ij}\sin\left(\theta_{ij}^k\right) - B_{ij}\cos\left(\theta_{ij}^k\right)\right)
\end{aligned} \tag{2.16}$$

- Real power flow measurement P_{flow} for the k-th iteration:

$$\begin{aligned}
\frac{\partial P_{ij}^{k+1}}{\partial \theta_i^{k+1}} &= V_i^k V_j^k\left(g_{ij}\sin\left(\theta_{ij}^k\right) - b_{ij}\cos\left(\theta_{ij}^k\right)\right) \\
\frac{\partial P_{ij}^{k+1}}{\partial \theta_j^{k+1}} &= -V_i^k V_j^k\left(g_{ij}\sin\left(\theta_{ij}^k\right) - b_{ij}\cos\left(\theta_{ij}^k\right)\right) \\
\frac{\partial P_{ij}^{k+1}}{\partial V_i^{k+1}} &= -V_j^k\left(g_{ij}\cos\left(\theta_{ij}^k\right) + b_{ij}\sin\left(\theta_{ij}^k\right)\right) + 2V_i^k\left(g_{ij} + g_{si}\right) \\
\frac{\partial P_{ij}^{k+1}}{\partial V_j^{k+1}} &= -V_i^k\left(g_{ij}\cos\left(\theta_{ij}^k\right) + b_{ij}\sin\left(\theta_{ij}^k\right)\right)
\end{aligned} \tag{2.17}$$

- Reactive power flow measurement Q_{flow} for the k-th iteration:

$$\begin{aligned}
\frac{\partial Q_{ij}^{k+1}}{\partial \theta_i^{k+1}} &= -V_i^k V_j^k\left(g_{ij}\cos\left(\theta_{ij}^k\right) + b_{ij}\sin\left(\theta_{ij}^k\right)\right) \\
\frac{\partial Q_{ij}^{k+1}}{\partial \theta_j^{k+1}} &= V_i^k V_j^k\left(g_{ij}\cos\left(\theta_{ij}^k\right) + b_{ij}\sin\left(\theta_{ij}^k\right)\right) \\
\frac{\partial Q_{ij}^{k+1}}{\partial V_i^{k+1}} &= -V_j^k\left(g_{ij}\sin\left(\theta_{ij}^k\right) - b_{ij}\cos\left(\theta_{ij}^k\right)\right) - 2V_i^k\left(b_{ij} + b_{si}\right) \\
\frac{\partial Q_{ij}^{k+1}}{\partial V_j^{k+1}} &= -V_i^k\left(g_{ij}\sin\left(\theta_{ij}^k\right) - b_{ij}\cos\left(\theta_{ij}^k\right)\right)
\end{aligned} \tag{2.18}$$

- Current measurement for the k-th iteration:

$$\begin{aligned}
\frac{\partial I_{ij}^{k+1}}{\partial \theta_i^{k+1}} &= \frac{g_{ij}^2 + b_{ij}^2}{I_{ij}^k} V_i^k V_j^k \mathrm{sen}\left(\theta_{ij}^k\right) \\
\frac{\partial I_{ij}^{k+1}}{\partial \theta_j^{k+1}} &= -\frac{g_{ij}^2 + b_{ij}^2}{I_{ij}^k} V_i^k V_j^k \mathrm{sen}\left(\theta_{ij}^k\right) \\
\frac{\partial I_{ij}^{k+1}}{\partial V_i^{k+1}} &= \frac{g_{ij}^2 + b_{ij}^2}{I_{ij}^k} \left(V_i^k - V_j^k \cos\left(\theta_{ij}^k\right)\right) \\
\frac{\partial I_{ij}^{k+1}}{\partial V_j^{k+1}} &= \frac{g_{ij}^2 + b_{ij}^2}{I_{ij}^k} \left(V_j^k - V_i^k \cos\left(\theta_{ij}^k\right)\right)
\end{aligned} \tag{2.19}$$

where:
$G_{ij} + jB_{ij}$: complex admittance at node ij
$g_{ij} + jb_{ij}$: series admittance at line connecting nodes i and j
$g_{si} + jb_{si}$: shunt complex admittance at line connected to node i

2.2.4 *Algorithm*

To proceed with the state estimation the following algorithm must be solved:

1. Initialize the iteration counter $k = 0$, and the states vector $x^T = [\theta_2\ \theta_3\ \ldots\ \theta_N\ V_1\ V_2\ \ldots\ V_N]$, where $V = 1$ *e* $\theta = 0$.
2. Calculate the measurements matrix $h(x)$.
3. Calculate the residuals matrix $R^k = z - h(x^k)$.
4. Calculate the Jacobian matrix H.
5. Calculate the Gain matrix G.
6. Calculate the right side of $[G(x^k)]\Delta x^{k+1} = H^T(x^k)W^{-1}R^k$.
7. Decompose $G(x^k)$ and to solve Δx^{k+1}.
8. Update the states vector $x^{k+1} = x^k + \Delta x^{k+1}$ and increment k.
9. Calculate the sum of weighted square residues, which is the objective function, by the formula $J^k = (R^k)^T W^{-1} R^k$.
10. Test convergence $\max|\Delta x^k| \leq$ threshold.
11. If converged, terminate the algorithm, otherwise return to step 2.

2.3 Didactic Example

For the numerical application of the state estimation technique, using the Classic WLS method, consider the 3-buses system below (Fig. 2.1), along with the line parameters (Table 2.1) and the measurement values (Table 2.2).

From the input data, the execution of the algorithm starts:

1. Assembly of the state vector with initial values $x^T = [0\quad 0\quad 1\quad 1\quad 1]$, where $\theta_0 = 0$ is the reference angle.

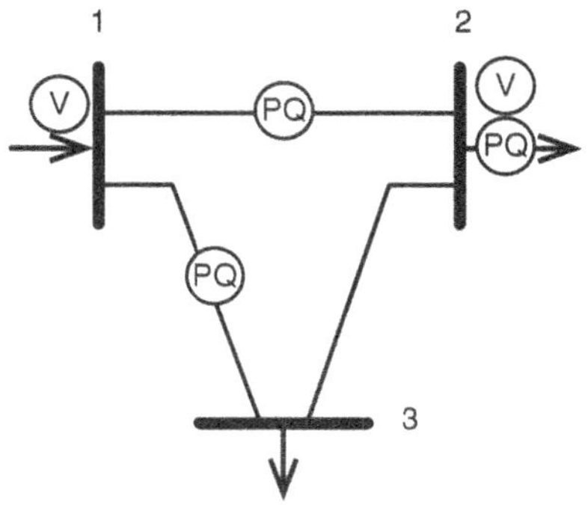

Fig. 2.1 three buses example system

Table 2.1 Line parameters

Line					
From bus	To bus	Resistance R (pu)	Reactance X (pu)	Susceptance B/2 (pu)	Tap Rel.
1	2	0.005	0.015	0.0	1
1	3	0.010	0.025	0.0	1
2	3	0.015	0.040	0.0	1

$S_{\text{base}} = 100$ MVA, $V_{\text{base}} = 230$ kV

Table 2.2 Admittance matrix

Node	1	2	3
1	−20.00 + j60.00	−13.79 + j34.48	
2	−20.00 + j60.00		−8.22 + j21.92
3	−13.79 + j34.48	−8.22 + j21.92	

Table 2.3 Measurements

Measurement	Type	Value (pu)	W
1	V_1	0.9910	1.6×10^{-5}
2	V_2	0.9620	1.6×10^{-5}
3	$P_{2,\text{inj}}$	−1.020	10^{-4}
4	$Q_{2,\text{inj}}$	−0.681	10^{-4}
5	$P_{12,\text{flow}}$	1.280	6.4×10^{-5}
6	$P_{13,\text{flow}}$	1.173	6.4×10^{-5}
7	$Q_{12,\text{flow}}$	0.863	6.4×10^{-5}
8	$Q_{13,\text{flow}}$	0.663	6.4×10^{-5}

2. Calculate the vector of measurement functions $h(x^k)$ through Eqs. (2.7–2.12). For this example, Table 2.4 shows the initial values as well as the calculated values in the algorithm iterations.
3. Calculate the residuals vector $R^k = z(x) - h(x^k)$. Table 2.5 shows the calculated values for the iterations of the algorithm.
4. Calculate the Jacobian matrix $H(x)$ of dimension $m \times n = 8 \times 5$ using Eqs. (2.14–2.19), in the following format:

Table 2.4 Measurements functions—$h(x^k)$

Measurement	Initial value	1st it.	2nd it.	3rd it.
1	1	0.986	0.9861	0.9861
2	1	0.967	0.9667	0.9667
3	0	−0.9849	−1.0225	−1.0226
4	0	−0.6727	−0.6893	−0.6893
5	0	1.23	1.2782	1.2782
6	0	1.1298	1.1741	1.174
7	0	0.8496	0.8561	0.8561
8	0	0.6833	0.6663	0.6663

Table 2.5 Residuals—$R^k = z(x) - h(x^k)$

Measurement	1st it.	2nd it.	3rd it.
1	0.005	0.0049	0.0049
2	−0.005	−0.0047	−0.0047
3	−0.0351	0.0025	0.0026
4	−0.0083	0.0083	0.0083
5	0.05	0.0018	0.0018
6	0.0432	−0.0011	−0.001
7	0.0134	0.0069	0.0069
8	−0.0203	−0.0033	−0.0033

$$H_{8x5} = \begin{bmatrix} 0 & 0 & \frac{\partial V_1}{\partial V_1} = 1 & 0 & 0 \\ 0 & 0 & 0 & \frac{\partial V_2}{\partial V_1} = 1 & 0 \\ \frac{\partial P_{2,\text{inj}}}{\partial \theta_2} & \frac{\partial P_{2,\text{inj}}}{\partial \theta_3} & \frac{\partial P_{2,\text{inj}}}{\partial V_1} & \frac{\partial P_{2,\text{inj}}}{\partial V_2} & \frac{\partial P_{2,\text{inj}}}{\partial V_3} \\ \frac{\partial Q_{2,\text{inj}}}{\partial \theta_2} & \frac{\partial Q_{2,\text{inj}}}{\partial \theta_3} & \frac{\partial Q_{2,\text{inj}}}{\partial V_1} & \frac{\partial Q_{2,\text{inj}}}{\partial V_2} & \frac{\partial Q_{2,\text{inj}}}{\partial V_3} \\ \frac{\partial P_{12,\text{flow}}}{\partial \theta_2} & \frac{\partial P_{12,\text{flow}}}{\partial \theta_3} & \frac{\partial P_{12,\text{flow}}}{\partial V_1} & \frac{\partial P_{12,\text{flow}}}{\partial V_2} & \frac{\partial P_{12,\text{flow}}}{\partial V_3} \\ \frac{\partial P_{13,\text{flow}}}{\partial \theta_2} & \frac{\partial P_{13,\text{flow}}}{\partial \theta_3} & \frac{\partial P_{13,\text{flow}}}{\partial V_1} & \frac{\partial P_{13,\text{flow}}}{\partial V_2} & \frac{\partial P_{13,\text{flow}}}{\partial V_3} \\ \frac{\partial Q_{12,\text{flow}}}{\partial \theta_2} & \frac{\partial Q_{12,\text{flow}}}{\partial \theta_3} & \frac{\partial Q_{12,\text{flow}}}{\partial V_1} & \frac{\partial Q_{12,\text{flow}}}{\partial V_2} & \frac{\partial Q_{12,\text{flow}}}{\partial V_3} \\ \frac{\partial Q_{13,\text{flow}}}{\partial \theta_2} & \frac{\partial Q_{13,\text{flow}}}{\partial \theta_3} & \frac{\partial Q_{13,\text{flow}}}{\partial V_1} & \frac{\partial Q_{13,\text{flow}}}{\partial V_2} & \frac{\partial Q_{13,\text{flow}}}{\partial V_3} \end{bmatrix}$$

For this example, the iterations k from 1 to 3 will have the Jacobian matrices $H(x)$ below:

$$H^{(k=1)} = \begin{bmatrix} 0 & 0 & 1 & 0 & 0 \\ 0 & 0 & 0 & 1 & 0 \\ 81.9178 & -21.9178 & -20 & 28.2192 & -8.2192 \\ -28.2192 & 8.2192 & -60 & 81.9178 & -21.9178 \\ -60 & 0 & 20 & -20 & 0 \\ 0 & -34.4828 & 13.7931 & 0 & -13.7931 \\ 20 & 0 & 60 & -60 & 0 \\ 0 & 13.7931 & 34.4828 & 0 & -34.4828 \end{bmatrix}$$

$$H^{(k=2)} = \begin{bmatrix} 0 & 0 & 1 & 0 & 0 \\ 0 & 0 & 0 & 1 & 0 \\ 77.2689 & -20.3516 & -20.2015 & 26.2686 & -7.7825 \\ -27.3709 & 7.4517 & -57.7239 & 78.5166 & -21.2552 \\ -57.4854 & 0 & 20.9679 & -18.8371 & 0 \\ 0 & -32.8425 & 14.7462 & 0 & -12.8258 \\ 18.215 & 0 & 60.0232 & -59.4488 & 0 \\ 0 & 12.2806 & 34.6939 & 0 & -34.3006 \end{bmatrix}$$

$$H^{(k=3)} = \begin{bmatrix} 0 & 0 & 1 & 0 & 0 \\ 0 & 0 & 0 & 1 & 0 \\ 77.2454 & -20.3529 & -20.2381 & 26.2223 & -7.7672 \\ -27.3947 & 7.4378 & -57.6941 & 78.4785 & -21.2544 \\ -57.4882 & 0 & 21.0183 & -18.7955 & 0 \\ 0 & -32.8649 & 14.7921 & 0 & -12.7805 \\ 18.1699 & 0 & 60.0346 & -59.4672 & 0 \\ 0 & 12.2384 & 34.6794 & 0 & -34.3206 \end{bmatrix}$$

5. Calculate the gain matrix $G(x)$ by Eq. (2.5). In this example, for iteration k from 1 to 3, matrices are as follows:

$$G^{(k=1)} = 10^7 \cdot \begin{bmatrix} 13.7568 & -2.0274 & 0.0548 & 0 & -0.0548 \\ -2.0274 & 2.7031 & -0.0548 & 0.0548 & 0 \\ 0.0548 & -0.0548 & 12.4114 & -11.7295 & -0.6757 \\ 0 & 0.0548 & -11.7295 & 13.7631 & -2.0274 \\ -0.0548 & 0 & -0.6757 & -2.0274 & 2.7031 \end{bmatrix}$$

$$G^{(k=2)} = 10^7 \cdot \begin{bmatrix} 12.4015 & -1.7765 & -0.156 & -0.1193 & -0.0196 \\ -1.7765 & 2.3907 & -0.11 & 0.0505 & 0 \\ -0.156 & -0.11 & 12.2832 & -11.2556 & -0.7708 \\ -0.1193 & 0.0505 & -11.2556 & 12.9377 & -1.8733 \\ -0.0196 & 0 & -0.7708 & -1.8733 & 2.6077 \end{bmatrix}$$

$$G^{(k=3)} = 10^7 \cdot \begin{bmatrix} 12.3971 & -1.7759 & -0.1663 & -0.1243 & -0.0177 \\ -1.7759 & 2.3912 & -0.1136 & 0.05 & 0 \\ -0.1663 & -0.1136 & 12.2872 & -11.254 & -0.7717 \\ -0.1243 & 0.05 & -11.254 & 12.9303 & -1.8717 \\ -0.0177 & 0 & -0.7717 & -1.8717 & 2.6078 \end{bmatrix}$$

6. Calculate the right side of Eq. (2.4) $[G(x^k)]\Delta x^{k+1} = H^T(x^k)W^{-1}R^k$
7. Decompose the matrix $G(x)$ for the calculation of Δx. For this example, in the iterations of k from 1 to 3, the obtained values of Δx are shown in Table 2.6.
8. Update the states vector by adding the value Δx^k and incrementing k. The updated states vector for iterations k from 1 to 3 are shown in Table 2.7.
9. Calculate the sum of weighted square residuals by the formula $J^k = (R^k)^T W^{-1} R^k$. For this example, the values of J^k for iterations k from 1 to 3 are in Table 2.8.
10. Make the convergence test $\max|\Delta x^k| \leq$ threshold. The threshold value used as the stopping criterion was 10^{-4}.
11. If convergence is reached, terminate the algorithm, otherwise return to step 2. In this example, convergence is obtained after iteration 3.

2.4 Observability and Ill-Conditioning

As already defined, a state estimator is nothing more than a telemetry measurement processor. The estimator receives a set of measurement values, each associated to a level of reliability, and knowing the physical characteristics of the power system, and

Table 2.6 Δx^k

State	1st it.	2nd it.	3rd it.
θ_2	−0.0149	−0.0007	2.3×10^{-7}
θ_3	−0.0227	−0.0013	4×10^{-8}
V_1	−0.014	0.0001	3.3×10^{-6}
V_2	−0.033	−0.0003	3.3×10^{-6}
V_3	−0.0425	0.0001	2.4×10^{-6}
$J(x^k)$	80,741	93.6679	4.6209

Table 2.7 State vector x^k

State	Initial value	1st it.	2nd it.	3rd it.
θ_2	0	−0.0149	−0.0156	−0.0156
θ_3	0	−0.0227	−0.024	−0.024
V_1	1	0.986	0.9861	0.9861
V_2	1	0.967	0.9667	0.9667
V_3	1	0.9575	0.9576	0.9576

Table 2.8 J^k

Iteration k	J^k
1	80,741.0
2	93.6679
3	4.6209

if these are correct, proposes a solution for each measure that minimizes, the square of the residuals weighted by the reliability levels. One question that can be made is: is it possible that any number of measures makes it possible to obtain a valid solution by a state estimator? The answer is no. A network that allows to obtain a valid solution is an observable network, and to know if a network is observable for the state estimation by a given set of measures observability analysis is necessary.

Observability analysis is a mandatory stage of state estimation, considering that telemetry is composed of physical elements, the telemetry system itself must be designed based on observability analysis. The observability analysis judges the quantity and quality of the measurements and should identify faults in the measurement system, as well as indicate measurement reinforcements that allow state estimation. These measurement reinforcements can be made by installing new meters with adequate accuracy as well as by pseudo-measures, which are measured from statistical processes. Pseudo-measures are measured with a lower degree of reliability and therefore should be used with caution, but they may be useful to achieve the level of observability in complex and poor monitored networks, as is the case with many distribution networks.

In addition to assisting in the design stage of the electric network monitoring, observability analysis can also be a runtime activity. Measurements happen in real time and communication instabilities, as well as many others, can change the amount set of online meters. In this way, it is the role of observability analysis: to verify in real time if the currently online meters are enough for the state estimation.

Thinking that the number of online meters, not always will be the totality projected, observability analysis allows to identify critical measurement points, which in case of unavailability, can make the whole system unobservable. Thus, the observability analysis may indicate possible measurement redundancies to increase the availability of the state estimator. Several works propose this in the literature [6–10]; however, we basically can typify the observability as numerical or topological.

- Numerical observability: the numerical observability is related to the rank of the gain matrix G. Since the matrix G is obtained through the Jacobian matrix H, then the matrix $H_{m \times n}$ must be observable, that is, it has minimally n linearly independent rows. In addition, the matrix G must be well conditioned, that is, it must be far from its singularity condition, which would prevent state estimation since the G matrix must be inverted [6]. The numerical observability allows to ensure the topological observability, but the calculation of this requires a great computational effort.
- Topological observability: unlike numerical observability, topological observability does not guarantee numerical observability. Despite this, it has enough practical usability to find problems in telemetry projects, thanks to the small computational effort required. In the problem modeling, graph theory is used [10], where the network is modeled as a graph containing the measurement scheme.

The problem of ill-conditioning in state estimation in power systems is similar to the observability problem in terms of its numerical consequence: in both cases the singularity of the G matrix makes its inversion impossible, as well as the calculation of the state vector. Just as in the case of observability a system can be well conditioned, but by an abnormal operation situation, the system can enter a state of ill-conditioning. It's a severe problem, since the system is ill-conditioned no extra effort in computing can produce an accurate result [11].

The problem is that, essentially, the solution of state estimation via Normal Equations is numerically sensitive/unstable, so several reasons for ill-conditioning can exist:

- Bad data in measurement values
- Quantity, type, and position of measurements
- High level of redundancy of measurements
- Large amount of power injection measurements, compared to the power flow measurements
- Characteristics of the network itself as the presence of short lines and the high R/X ratio, which is common in distribution networks [12]

The most effective solution to the problem of ill-conditioning is to mitigate the cause if this is possible. For example, in the case of bad data in measurement values it is possible to eliminate the measurement, verifying the observability, and to carry out the estimation without the measurement or with a pseudo-measurement.

For this reason, it is also very common to associate with state estimators also a real-time search for errors in the measurements, which can lead to ill-conditioning in some cases, as well as inaccurate results. Since the 1970s many studies have been developed with this objective [13–21], in particular, the largest normalized residual test method has good results in association with the WLS estimator, but an important vulnerability of this method is that it does not detect and identify multiple errors simultaneously [14]. Recently, methods to find multiple errors in measurements using phasor measurement units (PMUs) [17, 18] as well as other methods using topological and geometric approaches [19–21] have been proposed, instead of using state estimation residues [13–18].

2.5 Other State Estimation Techniques

The previously presented WLS state estimation technique was the first used in power systems since the 1970s and has been widely used. However, since the implementation of the first control centers using this technique, several deficiencies have been noticed, among them:

1. Numerical instability leads to inaccuracy in results.
2. Difficulty to identify bad data in more than one measurement.
3. It has a one-phase formulation, but many power systems, such as distribution networks, have imbalance between phases.
4. It does not allow to analyze the system in its dynamic aspect, neither to predict the next state, which could be useful for security analysis.
5. Its processing is complex, which becomes more evident as larger power systems are analyzed.
6. Processing is monolithic. Even if the power system has political and control divisions, the state estimator needs all the information at all times.

However, before discussing the solutions proposed in the literature for the mentioned problems it is possible to create classifications of the state estimators, based on the most relevant implementations already made. The first way to classify the estimators is given in temporal terms and the way the estimator captures the behavior of the power system over this time. Considering this, the state estimators can be classified as static and dynamic.

Another way to classify state estimators is in terms of how the results are computed. They can be computed in a monolithic way, considering the whole power system as a single network or, on the other hand, the power system can be divided into many areas with each area being the responsibility of a state estimator, this being an estimator called multi-area.

Next, the types of estimators for the two classifications will be defined.

2.5.1 Static State Estimators

Static state estimators were first named by Schweppe et al. [1] in the problem definitions and the WLS estimator itself. It is called static because of the very characteristics of the power system, which is a quasi-static system, since in normal situations changes are very slow, and therefore predictable, according to changes in loads. In this way the static estimators are the pioneers in the control centers, but that also have different implementations seeking to mitigate the deficiencies of the Classic WLS.

The problem of the numerical instability of the classic WLS method by Normal Equations, which often makes it impossible to find an accurate solution, can be treated in several ways. One of them is Orthogonal Factorization also known as factorization QR [22], which is more numerically robust and uses Givens rotations that are not computationally expensive. In this method the problem of normal equations is formulated differently, where the matrix G does not have to be obtained or factored. There is still the hybrid method that mixes the formulation by normal equations with the formulation by QR factorization [23]. Also on the factorization, it is possible to use the method of Peters and Wilkinson [24] that uses LU factorization in a formulation that combines the QR method with Normal Equations, but with a less ill-conditioned matrix than in the QR method.

An adopted practice in state estimation is the insertion of virtual measurements in some points of the network. Virtual measurements are measures that are fully certain of their value, are associated with a high level of reliability, and serve to strengthen the set of measures. However, a side effect of the insertion of measures with high reliability among measures of less reliability is that this leads to ill-conditioning of the G matrix and convergence problems. An alternative to this is the insertion of the virtual measures as explicit constraints in the WLS estimation [25]. The problem is solved by the Lagrangian method, where the weight matrix W has only small weights, avoiding one of the reasons for ill-conditioning.

Regarding the performance of the WLS method, approaches were developed with the premise of being more efficient, such as the decoupled approach using the same principle already used in decoupled power flow [26]. The fast decoupled state estimator [27] considers a constant approximation of the matrix G and H, keeping them constant during iterations. Since most of the processing in the WLS iteration comes from the computation of these matrices and factorization of the matrix G, in the decoupled algorithm this is done only in the first iteration, which drastically reduces the computational effort of the other iterations. However, this approximation made in the H and G matrices considers long lines and voltage levels above 69 kV, in addition to not considering current magnitude measurements, which limits the application basically to transmission systems.

The class of robust state estimators [13, 28, 29] was based on Huber's M-estimator [30], with the objective of being more robust than WLS when bad data occur. In addition, this method is more computationally efficient than WLS Classic, with execution times similar to the fast decoupled estimator [31]. These estimators are primarily designed to detect measures with a rapid growth of residues, reducing their influence on estimation during the iterative process. Subsequently, the sparse state recovery optimization method was proposed as a new class of robust m-estimators against measurement errors [32–34].

The estimators for distribution networks began to be studied in the mid-1990s [35–37], since then it was noticed that characteristics of these power systems made them incompatible with the previously used methods for the transmission systems. Among the characteristics that can be mentioned: imbalance between phases, high R/X ratio and mainly the poor density of measurements in view of the great complexity of these networks. In this way three-phase estimators were proposed [35–37], in addition to the use of pseudo-measurements, so that the networks could become observable for state estimation [38–40].

Although this class of estimators has emerged for more than two decades ago, the implementation of this by the utilities was very difficult due to the high cost involved in real-time monitoring and communication of the distribution networks. Nowadays, with the development of Smart Grids, the state estimators have played an important role since most of all real-time management and control applications will be based on the state of the network. Moreover, with the development of smart meters, state estimators can benefit from measurements with greater precision than pseudo-measurements, but there is a need to treat the synchronism problem in measurements [41–43] since the trend is that the intervals between measurements are long because of the large number of meters per network.

2.5.2 Dynamic State Estimators

Static state estimators have a very important role in the control centers when estimating the state vector from a set of measurements. Most of the time a power system can be considered quasi-static, since changes in loads and generations occur very slowly, which allows static state estimators to be executed at regular time intervals.

However, for the safety analysis of a power system more constant monitoring is required and even being able to predict the next state from the current state could save important time on control actions. An alternative would be to run a static state estimator at much shorter time intervals, but this would entail at least two major problems:

- Static estimators are computationally complex, and as networks and the set of measurements increase in complexity, they may not have a good response time for real-time monitoring purposes.
- Collecting a large set of measurements takes time, and doing this at very short intervals can cause non-synchronization between measurements, or even it may be impossible to do this.

To solve this problem, dynamic state estimators have been developed, which allow the evaluation of the power system continuously and also allow the prediction of the next state. The previously predicted state can be used as a pseudo-measurement in the next estimation, or to search for bad data. The first widely used dynamic state estimator was proposed in the 1970s [44], and subsequently improved [45]. Dynamic state estimators have been developed subsequently using the Extended Kalman Filter [45–47], which have been widely used.

In order to solve the problem of synchronism and precision of the obtained measurements, the phasor measurement units (PMU) were developed in early 1980s. PMUs allow the collection of magnitude and voltage angle data with great accuracy and in samplings much higher than the meters (at least 30 samples per second). In addition, each measurement receives a timestamp synchronized by the global positioning system (GPS), which allows to accurately compare voltage angles at different points in the network. PMUs were soon used in the dynamic state estimation, and because the measurement model was linear, simpler algorithms were used for estimation as the traditional Kalman Filter, which made the dynamic state estimation more precise and computationally efficient [48].

However, for economic reasons, it is not feasible to observe entire power systems only with PMUs, since they are expensive equipment and given the higher data generation also need a better communication infrastructure. Adding to the fact that there is already an entire infrastructure of conventional meters, a viable alternative is to mix the phasor measurements to the conventional magnitude measurements, creating hybrid static estimators [49–52].

A cheaper version of the PMUs for distribution networks was developed, the micro-PMUs [53]. These devices can also be applied for state estimation in a Smart Grid context [54–56].

2.5.3 *Single Area State Estimators*

All the previously mentioned state estimators, from which no multiple areas were mentioned, are monolithic or single area estimators.

2.5.4 *Multi-Area State Estimators*

The multi-area state estimators are estimators that allow the division of the estimation work between different areas, each one with its own estimator, and in the end the results are integrated to form a unique solution.

Studies for the multi-area state estimation have started in the 1970s [57–63] early in the state estimation in power systems. Since then, the goals with the multi-area state estimation basically were:

- Allow to integrate the results of different state estimators of interconnected systems, but subordinate to different control centers.

- Enable better response times, through distributed processing, in more complex power systems.

Multi-area state estimators are also known as multi-level [59] or hierarchical estimators, since different estimators can play the role of a local estimator or central estimator which is responsible for centralizing the results as well as coordinating the state estimation during iterations of the algorithm, depending on its implementation.

With the important role taken by state estimators in the implementation of Smart Grid concepts in distribution networks, there was also a need for the development of multi-area estimators for these networks [64–66]. Thanks to processing parallelism, that is possible with the multi-area estimators, it is possible to obtain adequate response times even with the great complexity imposed by these electrical networks.

Obviously, not all implementations of multi-area state estimators are static. The need for improvements in the measurement set and the accuracy of the solutions brought the PMUs and the dynamic state estimation also to multi-area or parallel implementations [34, 67, 68].

2.6 Conclusions

This chapter presents the state estimators general information and a review about other state estimators techniques. Evolutions in the distribution systems operation sector will require more aggregated information to enhance state estimation efficiency and accuracy. The distributed processing is necessary to accommodate the ever-increasing size of systems. Challenges are still open and new monitoring and decision-making solutions are required, such as accurate, adaptive and efficient feeder modeling and distribution systems state estimation methodologies.

References

1. Schweppe FC, Wildes J, Rom DB (1970) Power system static state estimation, part I, II and III. IEEE Trans Power Appar Syst PAS-89(1):120–135
2. Primadianto A, Lu CN (2017) A review on distribution system state estimation. IEEE Trans Power Syst 32(5):3875–3883
3. Wu FF (1990) Power system state estimation: a survey. Int J Electr Power Energy Syst 12 (2):80–87
4. Abur A, Exposito AG (2004) Power system state estimation: theory and implementation. CRC Press, Boca Raton, FL
5. Monticelli A (1999) State estimation in electric power systems: a generalized approach. Springer, New York
6. Krumpholz GR, Clements KA, Davis PW (1980) Power system observability: a practical algorithm using network topology. IEEE Trans Power Appar Syst PAS-99(4):1534–1542
7. Simões Costa AJA, Quintana VH, Mandel A (1982) Power system topological observability using direct graph-theoretic approach. IEEE Trans Power Appar Syst PAS(101):617–626

8. Monticelli A, Network Observability FFW (1985) Identification of observable Islands and measurement placement. IEEE Trans Power Appar Syst PAS(104):1035–1041
9. Monticelli A, Wu FF (1985) Network observability: theory. IEEE Trans Power Appar Syst PAS (104):1042–1048
10. Costa AS, Piazza TS, Mandel A (1990) Qualitative methods to solve qualitative problems in power system state estimation. IEEE Trans Power Syst 5(3):941–949
11. Rice JR (1981) Matrix computations and mathematical software. McGraw-Hill, New York
12. Abood HG, Sreeram V, Al-Saadi H, Mishra Y (2017) Solution methods of Ill-conditioned power system state estimation: a comparative study. IEEE Region 10 Symposium (TENSYMP)
13. Handschin E, Schweppe FC, Kohlas J, Fiechter A (1975) Bad data analysis for power system state estimation. IEEE Trans Power Appar Syst PAS-94(2):329–337
14. Cheniae MG, Mili L, Rousseeuw PJ (1996) Identification of multiple interacting bad data via power system decomposition. IEEE Trans Power Syst 11(3):1555–1563
15. Çelik MK, Abur A (1992) A robust WLAV state estimator using transformation. IEEE Trans Power Syst 7(1):106–113
16. Mili L, Phaniraj V, Rousseeuuw PJ (1991) Least median of squares estimation in power systems. IEEE Trans Power Syst 6(2):511–523
17. Chen J, Abur A (Nov. 2006) Placement of PMUs to enable bad data detection in state estimation. IEEE Trans Power Syst 21(4):1608–1615
18. Zhu J, Abur A (2007) Bad data identification when using phasor measurements. In Proc IEEE Power Tech Conf, Laussane, Switzerland, pp. 1676–1681
19. Clements KA, Davis PW (1986) Multiple bad data detectability and identifiably: a geometric approach. IEEE Trans Power Deliv 1(3):355–360
20. Bretas NG, Bretas AS (2011) Innovation concept for measurement gross error detection and identification in power system state estimation. IET Gener Transm Distrib 5:603–608
21. Bretas NG, Bretas AS, Martins ACP (2013) Convergence property of the measurement gross error correction in power system state estimation, using geometrical background. IEEE Trans Power Syst 28(4):3729–3736
22. Simões-Costa AJA, Quintana VH (1981) A robust numerical technique for power system state estimation. IEEE Trans Power Appar Syst PAS-100:691–698
23. Monticelli A, Murari C, A Hybrid State Estimator FW (1985) Solving normal equations by orthogonal transformations. IEEE Trans Power Appar Syst PAS-105(2):3460–3468
24. Gu J, Clements K, Krumpholz G, Davis P (1983) The solution of ill-conditioned power system state estimation problems via the method of Peters and Wikinson. PICA Conference Proceedings, Houston, pp. 239–246
25. Aschmoneit F, Peterson N, Adrian E (1977) State estimation with equality constraints. 10th PICA Conference Proceedings, Toronto, pp. 427–430
26. Stott B, Alsac O (1974) Fast decoupled load flow. IEEE Trans Power Appar Syst 93:859–869
27. Monticelli A, Garcia A (1990) Fast decoupled state estimators. IEEE Trans Power Syst 5 (2):556–564
28. Merrill HM, Schweppe FC (1971) Bad data suppression in power system static state estimation. IEEE Trans Power Appar Syst PAS-90:2718–2725
29. Mili L, Cheniae MG, Vichare NS, Rousseeuw PJ (1996) Robust state estimation based on projection statistics. IEEE Trans Power Syst 11(2):1118–1127
30. Huber PJ (1964) Robust estimation of a location parameter. Ann Math Stat 35:73–101
31. Wu W, Guo Y, Zhang B, Bose A, Hongbin S (2011) Robust state estimation method based on maximum exponential square. IET Gener Transm Distrib 5(11):1165–1172
32. Xu W, Wang M, Cai J, Tang A (2013) Sparse error correction from nonlinear measurements with applications in bad data detection for power networks. IEEE Trans Signal Process 61 (24):6175–6187
33. Kekatos V, Giannakis GB (2013) Distributed robust power system state estimation. IEEE Trans Power Syst 28(2):1617–1626

34. Zheng W, Wu W, Gomez-Exposito A, Zhang B, Guo Y (2017) Distributed robust bilinear state estimation for power systems with nonlinear measurements. IEEE Trans Power Syst 32 (1):499–509
35. Roytelman I, Shahidehpour SM (1993) State estimation for electric power distribution systems in quasy real time conditions. IEEE Trans Power Delivery 8(4):2009–2015
36. Baran ME, Kelley AW (1994) State estimation for real time monitoring of distribution system. IEEE Trans Power Syst 9(3):1601–1609
37. Baran ME, Kelley AW (1995) A branch current based state estimation method for distribution systems. IEEE Trans Power Syst 10(1):483–491
38. Ghosh AK, Lubkeman DL, Jones RH (1997) Load modeling for distribution circuit state estimation. IEEE Trans Power Delivery 12(2):999–1005
39. Manitsas E, Singh R, Pal B, Strbac G (2008) Modelling of pseudo measurements for distribution system state estimation. Proc. Int. Conf. Exhib. Elect. Distrib. Seminar: SmartGrids Distrib., Frankfurt, Germany, pp. 1–4
40. Singh R, Pal BC, Jabr RA (2010) Distribution system state estimation through gaussian mixture model of the load as pseudo-measurement. IET Gener Transm Distrib 4(1):50–59
41. Hayes BP, Gruber JK, Prodanovic M (2014) A closed-loop state estimation tool for MV network monitoring and operation. IEEE Trans Smart Grid 6(4):2116–2125
42. Exposito AG, Quiles CG, Dzafic I (2015) State estimation in two time scales for smart distribution systems. IEEE Trans Smart Grid 6(1):421–430
43. Alimardani A, Therrien F, Atanackovic D, Jatskevich J, Vaahedi E (2015) Distribution system state estimation based on nonsynchronized smart meters. IEEE Trans Smart Grid 6 (6):2919–2928
44. Debs A, Larson R (1970) A dynamic estimator for tracking the state of a power system. IEEE Trans Power Appar Syst PAS-89(7):1670–1678
45. Silva AL, Coutto Filho MB, Queiroz J (1983) State forecasting in electric power systems. IEE Gene Trans Distrib 130(5):237–244
46. Silva AM, Coutto Filho M, Cantera J (1987) An efficient dynamic state estimation algorithm including bad data processing. IEEE Trans Power Syst 2(4):1050–1058
47. Ghahremani E, Kamwa I (2011) Dynamic state estimation in power system by applying the extended Kalman filter with unknown inputs to phasor measurements. IEEE Trans Power Syst 26(4):2556–2566
48. Sarri S, Zanni L, Popovic M et al (2016) Performance assessment of linear state estimators using synchrophasor measurements. IEEE Trans Sustain Energy 65(3):535–548
49. Aminifar F, Shahidehpour M, Fotuhi-Firuzabad M et al (2014) Power system dynamic state estimation with synchronized phasor measurements. IEEE Trans Instrum Meas 63(2):352–363
50. Chakhchoukh Y, Vittal V, Heydt GT (2014) PMU based state estimation by integrating correlation. IEEE Trans Power Syst 29(2):617–626
51. Du J, Ma S, Wu YC et al (2014) Distributed hybrid power state estimation under PMU sampling phase errors. IEEE Trans Signal Process 62(16):4052–4063
52. Tavares B, Freitas V, Miranda V, Costa AS (2017) Merging conventional and phasor measurements in state estimation: a multi-criteria perspective. 19th Int. Conf. on Intelligent System Application to Power Systems (ISAP)
53. Von Meier A, Culler D, McEachern A, Arghandeh R (2014) Micro-synchrophasors for distribution systems. In: Proceedings of the Innov. Smart grid Technol. Conference (ISGT), IEEE PES, pp. 1–5
54. Göl M, Abur A (2014) A robust PMU based three-phase state estimator using modal decoupling. IEEE Trans Power Syst 29(5):2292–2299
55. Sarri S, Paolone M, Cherkaoui R, et al. (2012) State estimation of active distribution networks: comparison between WLS and iterated Kalman-filter algorithm integrating PMUs. Innov Smart Grid Technol ISGT Eur 2012, Proc of the 3rd IEEE PES International Conference Exhib, pp. 1–8

56. Haughton DA, Heydt GT (2013) A linear state estimation formulation for smart distribution systems. IEEE Trans Power Syst 28:1187–1195
57. Clements KA, Denison OJ, Ringlee RJ (1972) A multi-area approach to state estimation in power system networks. IEEE PES Summer Meeting, Paper C72 465-3, San Francisco, CA
58. Kobayashi H, Narita S, Hamman MSSA (1974) Model coordination method applied to power system control and estimation problems. Proceedings of the IFAC/IFIC 4th International Conference on Digital Computer Appl to Power Control, pp. 114–128
59. Van Cutsem T, Ribbens-Pavella M (1981) A two-level static state estimator for electric power systems. IEEE Trans Power Appar Syst PAS-100(8):3722–3732
60. Conejo AJ, de la Torre S, Canas M (2007) An optimization approach to multiarea state estimation. IEEE Trans Power Syst 22(1):213–221
61. Korres GN (2011) A distributed multiarea state estimation. IEEE Trans Power Syst 26(1):73–84
62. Xie L, Choi D-H, Kar S, Poor HV (2012) Fully distributed state estimation for wide-area monitoring systems. IEEE Trans Smart Grid 3(3):1154–1169
63. Korres GN, Tzavellas A, Galinas E (2013) A distributed implementation of multi-area power system state estimation on a cluster of computers. Electr Pow Syst Res 102:20–32
64. Nusrat N, Irving M, Taylor G (2011) Development of distributed state estimation methods to enable smart distribution management systems. IEEE Int Symp Ind Electr (ISIE):1691–1696
65. Garcia LDA, Grenard S (2011) Scalable distribution state estimation approach for distribution management systems. Proc. 2nd IEEE PES Int. Conf. Exhibit. Innov. Smart Grid Technol. (ISGT Europe), pp. 1–6
66. Muscas C, Pau M, Pegoraro PA, Sulis S (2015) Multiarea distribution system state estimation. IEEE Trans Instr Meas 64(5):1140–1148
67. Qing X, Karimi H, Niu Y, Wang X (2015) Decentralized unscented Kalman filter based on a consensus algorithm for multi-area dynamic state estimation in power systems. Int J Electr Power Energy Syst 65:26–33
68. Huang R, Cokkinides G, Hedrington C (2016) Distribution system distributed quasi-dynamic state estimator. IEEE Trans Smart Grid 7(6):2761–2770

Chapter 3
Self-Healing in Power Distribution Systems

M. J. S. Ramos, M. Resener, P. H. E. Oliveira, and Daniel Pinheiro Bernardon

3.1 Introduction

The electric energy supply continuity is a constant goal in electrical utility companies. This goal represents not only comfort and satisfaction for customers, but also an ever-increasing need for commercial and industrial activities, as interruptions compromise production. However, in order to execute the expansion of the networks or preventive and corrective maintenance actions, interruptions are usually needed. Furthermore, when a fault occurs in the system, the protection devices isolate the faulted portion of the network, interrupting the electric energy supply, which places the network in the condition of contingency [1].

In cases where power grid failures and consequent contingency occur, the restoration of the electric energy supply, also called energy restoration, is of great importance to the utilities, which have continuously applied efforts to make the process and thus reduce the impact of energy interruption on consumers.

M. J. S. Ramos (✉) · D. P. Bernardon
Technology Center, Federal University of Santa Maria, Santa Maria, Rio Grande do Sul, Brazil
e-mail: dpbernardon@ufsm.br

M. Resener
Department of Electrical Systems of Automation and Energy, Federal University of Rio Grande do Sul, Porto Alegre, Rio Grande do Sul, Brazil
e-mail: mariana.resener@ufrgs.br

P. H. E. Oliveira
RGE Sul Power Utility, São Leopoldo, Rio Grande Sul, Brazil
e-mail: paulo.henriquee@rgesul.com.br

D. P. Bernardon, V. J. Garcia (eds.), *Smart Operation for Power Distribution Systems*, https://doi.org/10.1007/978-3-319-93922-3_3

Considering the emergence of new technologies and constant evolution, automated distribution equipment and systems for communication and control are increasingly common in power grids and in Operation Centers. Currently, efforts are related to the application models for automatic restoration of electricity supply, and there are different application models for each company. Energy management and self-healing are important features of a smart distribution system. Since the self-healing task is usually performed under critical conditions after faults, the operational and time constraints can add to the complexity of the problem [2, 3].

With the ever-increasing penetration levels of distributed generators (DGs), the stochasticity of non-dispatchable DGs and loads brings new challenges to the operation and self-healing of distribution systems [4].

This chapter describes the main steps of the process related to the restoration of the electric energy supply in distribution systems.

3.2 General Explanation

Considering the electrical network in a contingency condition, it is desired the faulted area be identified and isolated quickly, restricting the out-of-service area to a minimum. In general, when a fault occurs at any point in the network, the following actions must be taken [5]:

- Identify the location where the fault occurred.
- Isolate the smallest possible part of the system by switching protection devices and/or maneuvering equipment.
- Switch equipment to restore the supply to the out-of-service consumers upstream and/or downstream of the isolated network segment.
- Act to eliminate the fault of the network.
- Perform new switchings to return feeders involved to the normal operating state topology.

Among the solutions to achieve adequate power restoration times and to meet performance limits, one that has been increasingly used by utilities is the use of automation systems in distribution networks. In this case, one can cite the application of automated reclosers and/or remotely operated switches. These devices have shown economically viable due to the emergence of several automation equipment suppliers and new communication link technologies [6].

In addition to the technologies for remote operation, the concepts of self-healing and smart grid can be applied in energy restoration. These concepts allow the automation of the distribution networks, automatically performing the operations in the remote-controlled equipment, such as the isolation of faults and load transfers, resulting in reliable actions and significantly reducing the energy supply restoration time [7, 8].

The use of an efficient methodology for the operation of these equipments is important for utilities, in order to guarantee the technical feasibility of the switchings and to optimize the restoration process, reducing the time of energy restoration.

The development of a suitable and universal method for analyzing the problem in question is difficult, given the number of factors to be considered. Thus, one can achieve more efficient solutions when the problem is solved for each utility, individually, considering its distribution network structure, quantity, and types of protection and switching equipment, systems for supervision and control of equipment, and adequacy of database, among others [9].

The problem related to the automatic reestablishment of electric power distribution networks can be basically divided into two operating philosophies: (1) decentralized system and (2) centralized system. The concepts related to both philosophies will be discussed further in this chapter.

Another aspect to consider is the post-contingency network operation in an island state, being part of the network supplied by a distributed generation source. This issue is still incipient for distribution systems in most countries, and the current regulation do not specifically address this. In fact, the use of protection systems with anti-islanding scheme is required in most cases, mainly due to operational safety issues. However, with the prospect of a significant increase in distributed generation sources, island-mode operation after contingency becomes attractive for reliability purposes [10–12]. For rural systems, for instance, which have few interconnections between feeders, the islanded mode is often the only option for alternative energy supply. Indeed, further studies are needed in this area, including several technical issues related to automation and control of loads and generators. In this context, this chapter discusses the possibility of islanded operation in contingency situations, in order to show the improvement in reliability.

3.3 Location of the Faulty Area

The location of the fault is a very important step in the process of restoring the electric power supply. In practice, the protection philosophy adopted by each utility has a direct impact on the model used to recompose the power grid. It is important to understand that the protection philosophy guides the reconfiguration of the network by Operators in Operation Centers, which follow procedures based on the interpretation of the information obtained by the automated equipment installed along the distribution network and/or in the substation, besides the information of the customers affected by the electric energy supply interruption.

Currently, many electric power utilities have a large amount of automated equipment along the networks, so that Operators of Operation Centers carry out quick operations for restoring the supply, based on the information from the field equipment. This information can be obtained with a Supervisory Control and Data Acquisition (SCADA) system, as well as all the commands.

The central idea of the concept presented here is based on applying in Self-Healing systems the same interpretations that are used by operators to restore the power supply. Such an approach ensures that the levels of security applied by Operators will be maintained by the automated system, even in cases where communication failure, protection mismatch, simultaneous failures, and hotline maintenance occur, among other possibilities.

3.3.1 Definition of Clear and Protected Branches

The concept of clear and protected branches is based on the information of the protections signaled from the field equipment, where it is possible to identify the short-circuit current passing by the automated protection and/or switching equipment, besides the tripping information of protective equipment. Also included in this item are the information related to manual protection devices, as fuse keys registered in the Distribution Management System (DMS).

The definition of clear branches is related to the portion of the distribution network where it is possible to carry out the energization without the need for visual inspection, since the information obtained by the automated field equipment does not indicate any anomaly in the respective branches. On the other hand, the protected branch is characterized by the need for visual inspection, since the automated field equipment or fuses registered in the DMS indicate the possibility of failure in the respective branches of the network.

Figures 3.1 and 3.2 illustrate the clear and protected branches of a medium voltage feeder under fault. The branches marked in green refer to clear branches, while the branches marked in red are those protected. It is important to emphasize that the definition of the sections is based on the utility's protection philosophy, the field equipment information, and the DMS data.

The clear and protected branches approach can be easily applied in a centralized self-healing system, where all the data are integrated through an Advanced Distribution Management System (ADMS). On the other hand, for a decentralized self-healing system, this approach must be modified due to limitation in the information sent to the DMS system by the field equipment.

Fig. 3.1 Feeder overview

Fig. 3.2 View of clear and protected branches

3.3.2 Network Topology from Signaled Protections

Power distribution networks have different types of protection equipment, which have different protection functions. Because of this diversity, one can obtain different network topologies from the protection functions signaled in the automated equipment or in the substation breaker. The definition of these topologies, presented as follows, allows a better understanding of the clear and protected branches, as well as a better understanding of the operation of distribution systems.

3.3.2.1 Topology N° 1: Ideal Network (Generic Condition)

Power distribution networks have different types of equipment. In the ideal network topology, the protection devices are perfectly selective, that is, the protective equipment only acts for faults in the system. There is an Automatic Switch (ASC) able to identify short-circuit currents. This condition is greatly affected by the short-circuit levels of the system, since the coordination and selectivity of protective equipment becomes more complex as short-circuit levels increase. Figures 3.3 and 3.4 illustrate an ideal system and the time coordination curves, respectively. It can be seen that all protection devices are perfectly selective, which in practice is not feasible in distribution systems with high levels of short-circuit current.

3.3.2.2 Topology N° 2: High Short-Circuit Current

The second topology is characterized with high short-circuit levels. In this case, 100% selectivity cannot be guaranteed due to instantaneous protection settings. Figure 3.5 presents the network topology where it is possible to identify the clear and protected branches based on the alarms signaled in the SCADA and shown in Fig. 3.6. In Fig. 3.7 one can observe the high short-circuit current value tripping all protective equipment.

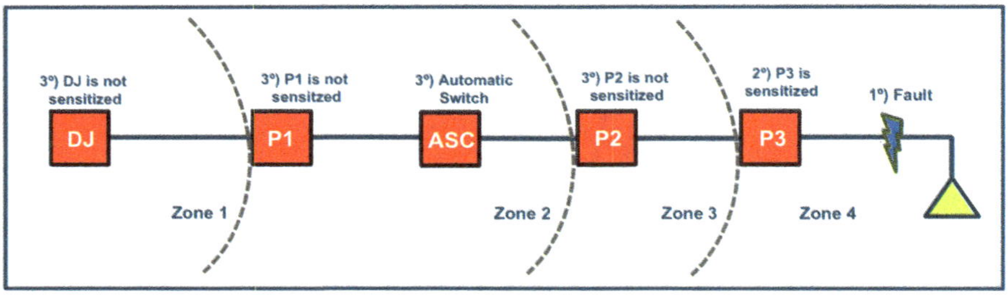

Fig. 3.3 Topology N° 1, ideal network (100% of selectivity)

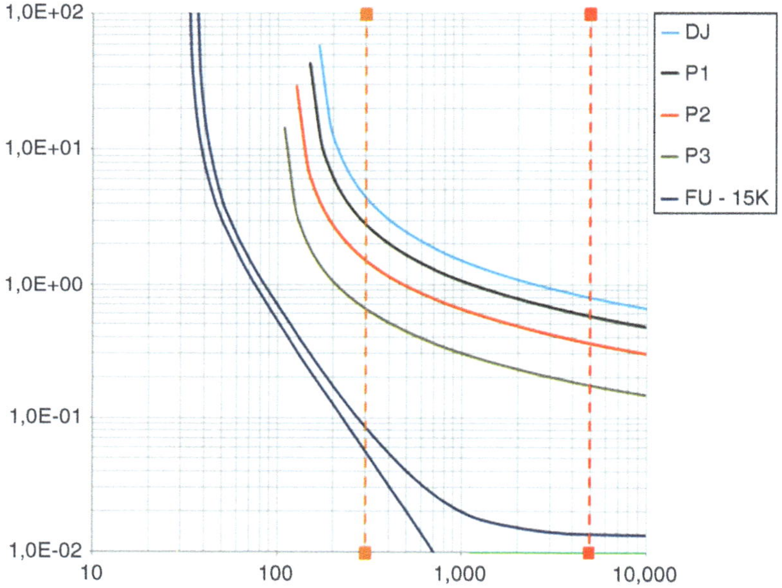

Fig. 3.4 Time coordination curves (100% of selectivity)

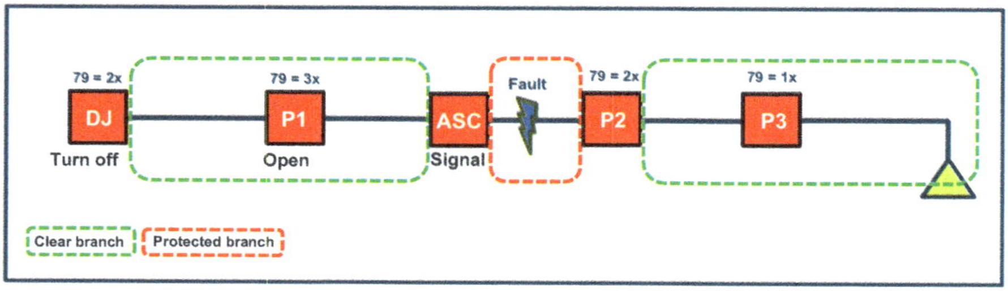

Fig. 3.5 Topology N° 2, high short-circuit current values

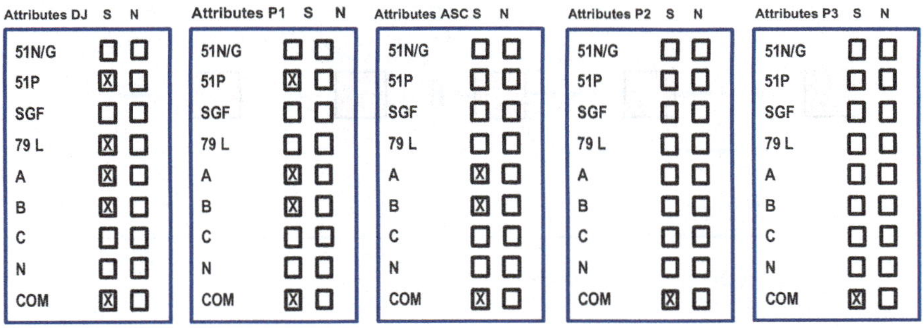

Fig. 3.6 Attributes for identification of the equipment signaled in SCADA—Topology N° 2

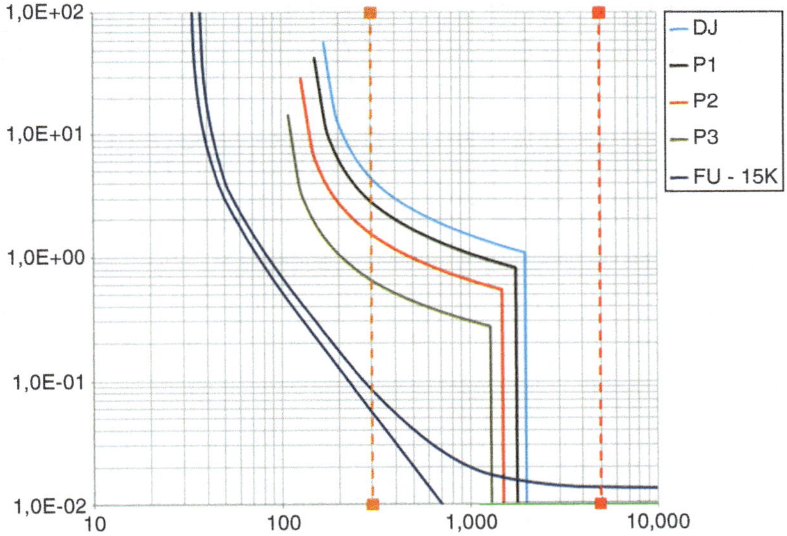

Fig. 3.7 Time coordination curves—Topology N° 2

3.3.2.3 Topology N° 3: Low Short-Circuit Current

In this topology the short-circuit currents are low, so it is possible to obtain selectivity between the protection devices as the current levels reduce significantly due to fault impedance and/or the distribution network. Figure 3.8 illustrates the clear and protected branches based on the alarms signaled in the SCADA, according to Fig. 3.9. In Fig. 3.10 one can observe that for low values of short-circuit current, the system becomes 100% selective, i.e., it only switches off the equipment upstream of the fault.

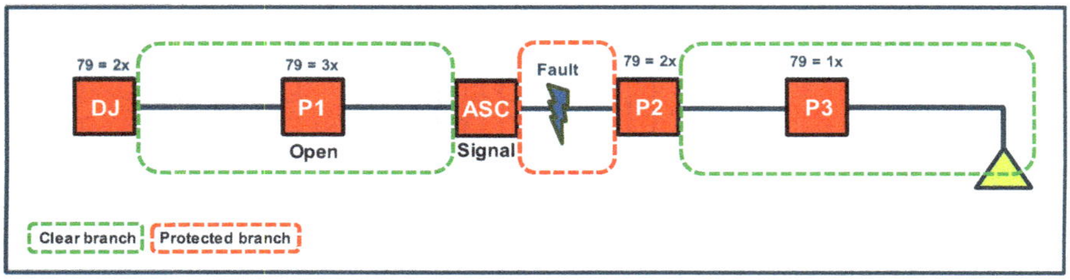

Fig. 3.8 Topology N° 3, low short-circuit currents

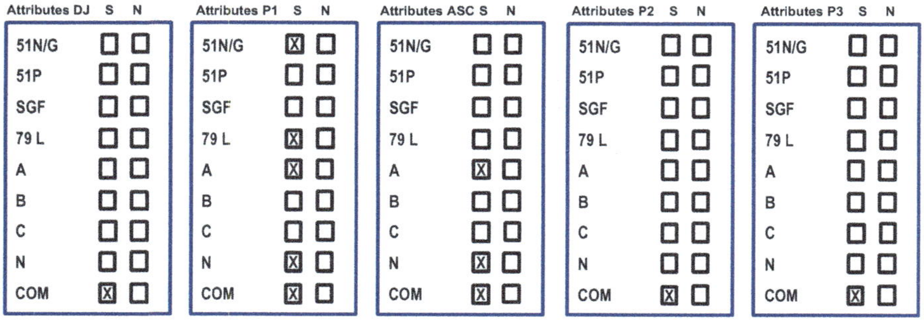

Fig. 3.9 Attributes for identification of the equipment signaled in SCADA—Topology N° 3

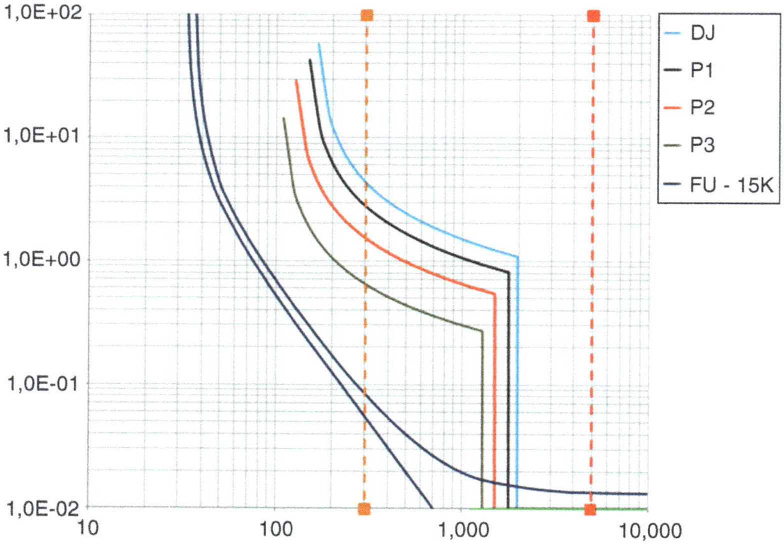

Fig. 3.10 Time coordination curves—Topology N° 3

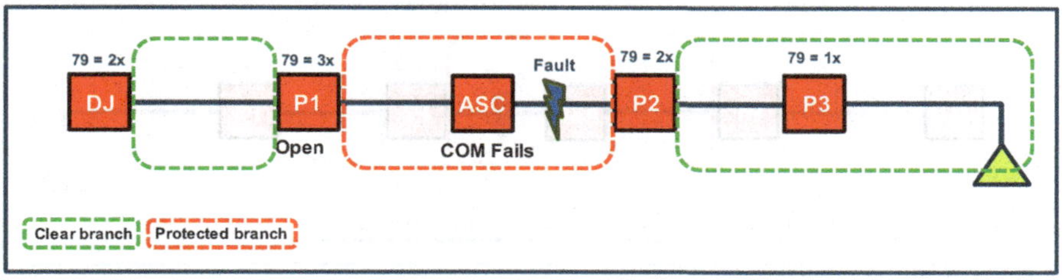

Fig. 3.11 Topology N° 4, communication failure

Attributes	DJ S	DJ N	P1 S	P1 N	ASC S	ASC N	P2 S	P2 N	P3 S	P3 N
51N/G	☐	☐	☒	☐	☐	☐	☐	☐	☐	☐
51P	☐	☐	☐	☐	☐	☐	☐	☐	☐	☐
SGF	☐	☐	☐	☐	☐	☐	☐	☐	☐	☐
79 L	☐	☐	☒	☐	☐	☐	☐	☐	☐	☐
A	☐	☐	☒	☐	☐	☐	☐	☐	☐	☐
B	☐	☐	☐	☐	☐	☐	☐	☐	☐	☐
C	☐	☐	☐	☐	☐	☐	☐	☐	☐	☐
N	☐	☐	☒	☐	☐	☐	☐	☐	☐	☐
COM	☒	☐	☒	☐	☐	☒	☒	☐	☒	☐

Fig. 3.12 Attributes for identification of the equipment signaled in SCADA—Topology N° 4

3.3.2.4 Topology N° 4: Communication Failure

The example presented in the following can be applied in both low and high fault current conditions. In this topology, the failure of communication in some automated equipment can completely change the system restoration.

Figure 3.11 presents a topology of a network with clear and protected sections, based on the alarms signaled in the SCADA, and illustrates the "ASC" equipment with communication failure, which cannot be considered in the solution. One can see in this case the protected area increased in relation to that shown in Fig. 3.8. Figure 3.12 shows the attributes of the reclosers that are part of the solution. It is possible to notice that the equipment "ASC" presents a failure in the communication and no signaling is verified for the fault in the distribution network. The time coordination curve of this system is the same as shown in Fig. 3.10.

3.3.2.5 Topology N° 5: Sensitive Ground Fault Performance Between Reclosers and Fuses

The fifth topology has a system with very low short-circuit currents, so it is not possible to de-energize the distribution network through traditional protection functions (51P and 51N/G). Currently, an alternative available in digital protection equipment is the sensitive ground fault (SGF) or 51GS function, which acts for

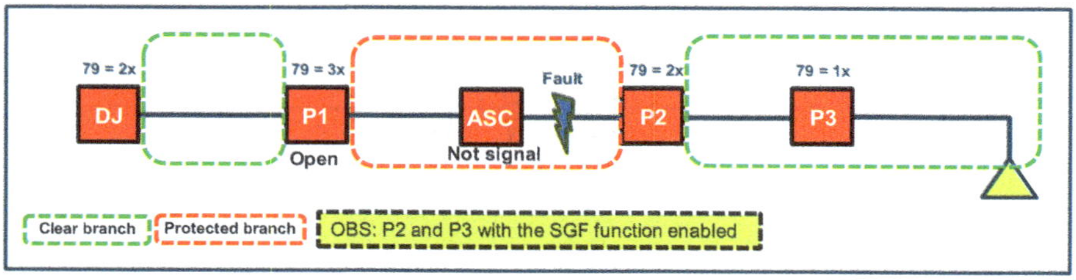

Fig. 3.13 Topology N° 5-A, SGF function signaling between reclosers (all reclosers with this function enabled)

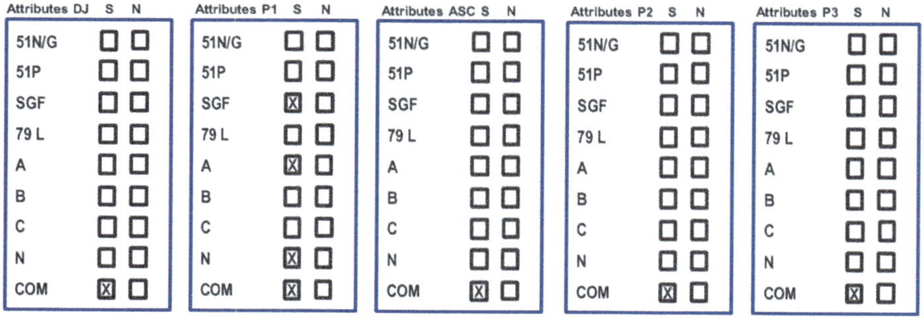

Fig. 3.14 Attributes for identification of the equipment signaled in SCADA—Topology N° 5-A

very small values of current and with a long time of operation. The example below considers SGF available on all network reclosers.

Figure 3.13 shows the clear and protected branches after the SGF function has actuated on P2. Figure 3.14 shows that the equipment "ASC" and "P1" did not signalized, and for the "P1" device there is a timing for selectivity between the SGF of P1 and P2. Figure 3.15 shows the time coordination curves for this case.

Figure 3.16 shows the clear and protected branches in the presence of fuses upstream of the fault. In this example, all the reclosers have the SGF function enabled and it is possible to verify that for faults downstream of the 25 K fuses the SGF function is sensitized. The fuses to be considered as protected branches will vary according to the protection philosophy of each utility. Figure 3.17 presents the attributes for identification of the equipment signaled in the SCADA. The time coordination curves of this system are presented in Fig. 3.15.

3.3.2.6 Topology N° 6: SGF Function Enabled in the Backup Recloser

This topology is characterized by very low short-circuit currents, where it is not possible to de-energize the distribution network through traditional protections (51P and 51N/G). The SGF function presents as an alternative available in digital devices, which acts for very small values of current with a time delay. The example below considers the function enabled only on device P1.

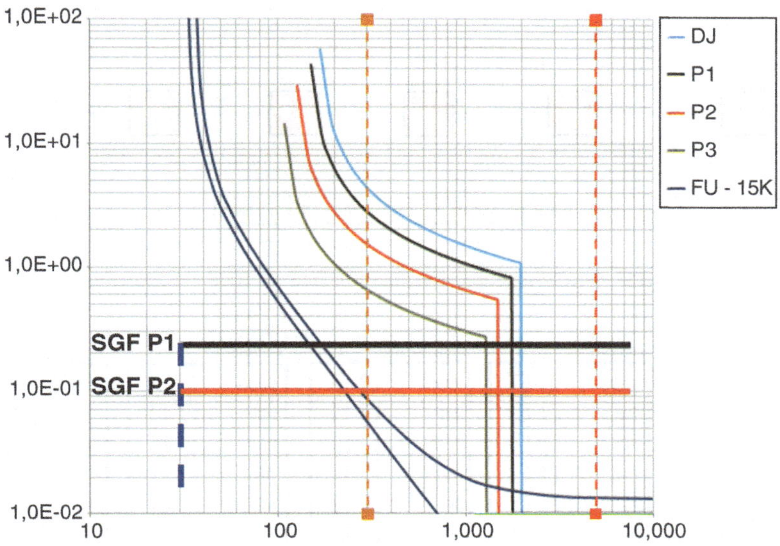

Fig. 3.15 Time coordination curves—Topology N° 5-A

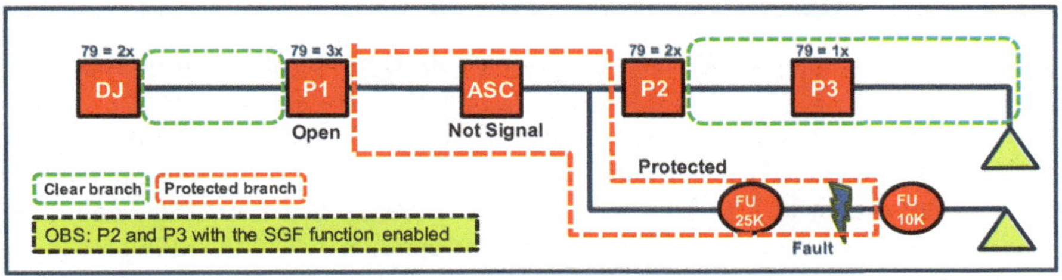

Fig. 3.16 Topology N° 5-B, SGF function signaling between reclosers

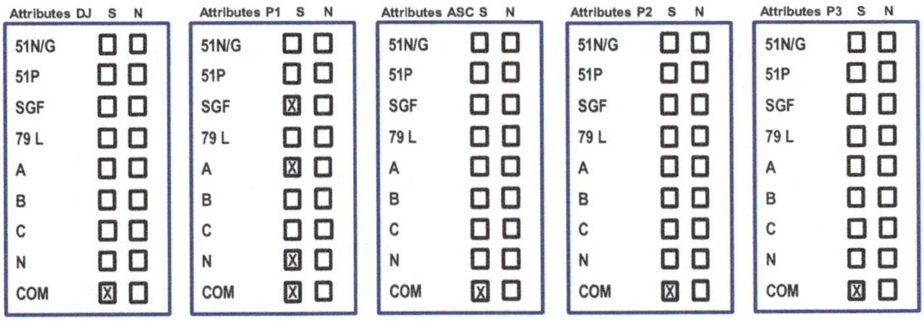

Fig. 3.17 Attributes for identification of the equipment signaled in SCADA—Topology N° 5-B

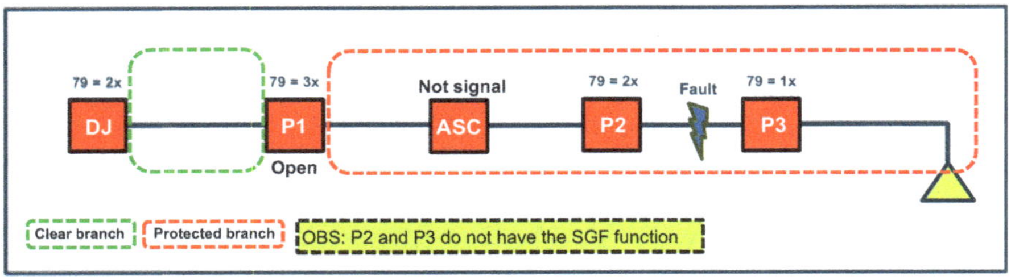

Fig. 3.18 Topology N° 6, SGF function signalized only on the backup recloser

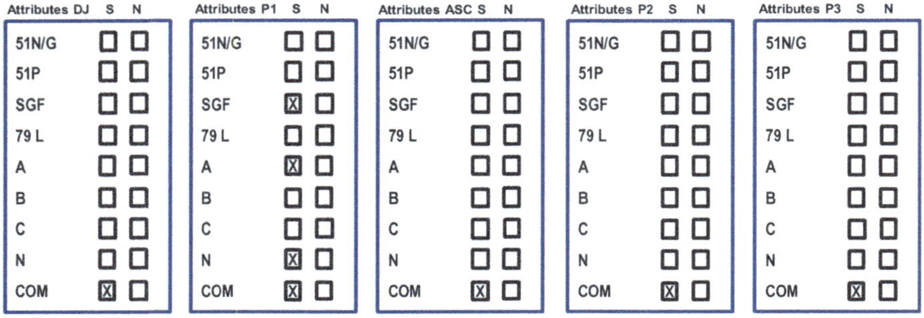

Fig. 3.19 Attributes for identification of the equipment signaled in SCADA—Topology N° 6

Figure 3.18 shows the clear and protected branches based on the alarms signaled in the SCADA system. The attributes of the automated equipment that are part of the solution are shown in Fig. 3.19. It is possible to verify that the "ASC" and "P2" equipment do not signal for faults with very low currents, where the SGF protection is sensitized. Figure 3.20 illustrates the time coordination curves of this system topology.

3.3.2.7 Topology N° 7: Uncoordinated System Due to Maneuvered Network

This topology is the most critical condition from the operational point of view, where the distribution network configuration is modified due to the need for emergency load transfers. Such need is quite common in the day-to-day life of an electric power distribution company. In many situations in which the network configuration is changed, the protection system may not be selective, since it is not designed for such topology of the feeder. Figure 3.21 shows the topology of the network in its steady-state operating condition, while Fig. 3.22 shows the time coordination curves of this system.

Figure 3.23 shows the topology of the network in its operating condition after the transfer of emergency loads has occurred, while Fig. 3.24 shows the attributes for

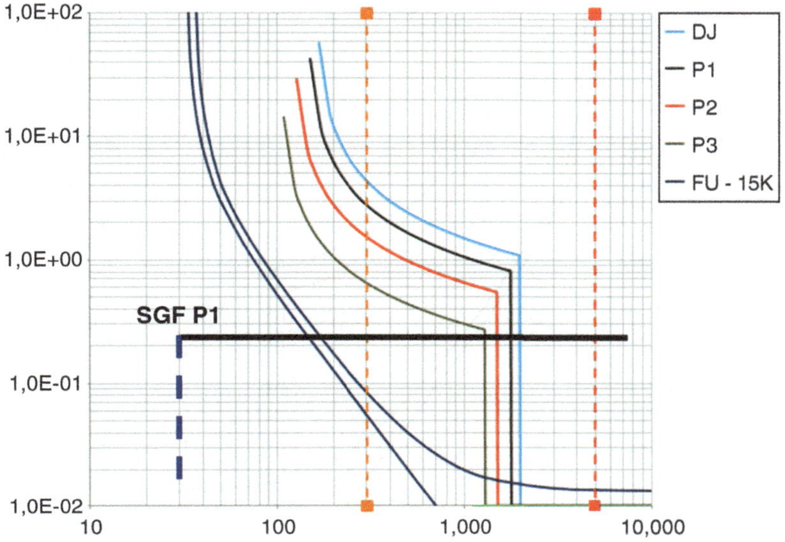

Fig. 3.20 Time coordination curves—Topology N° 6

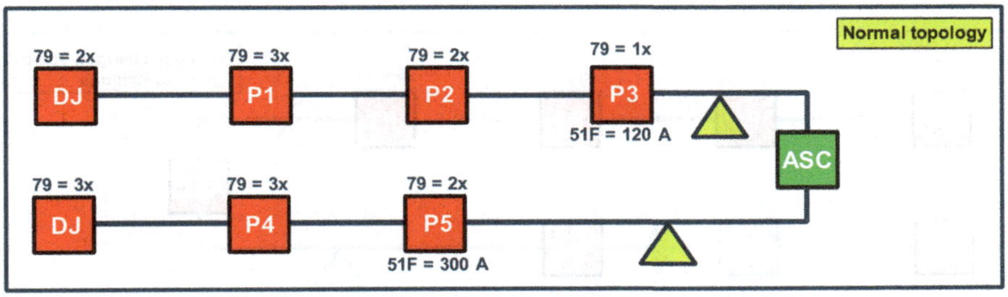

Fig. 3.21 Topology N° 7, network in the steady-state operating condition

identification of the equipment signaled in SCADA. In Fig. 3.25 the time coordination curves are presented for this topology, where it is possible to verify there is an uncoordinated protection between reclosers "P3" and "P5."

In addition to the attributes presented in Figs. 3.24 and 3.25 for the cases where the network had its configuration changed, the short-circuit currents must also be analyzed in order to identify the fault location. In what follows, a logic to identify the location of the fault based on the currents recorded by the automated equipment along the distribution network is described:

$$(\text{IF } Icc_{P3} > 51\text{P}_{P5}) \text{ AND } (\text{P5}_{\text{NOT TRIP}}), \text{fault location between P3 and P5} \quad (3.1)$$

$$(\text{IF } Icc_{P3} < 51\text{P}_{P5}) \text{ AND } (\text{P5}_{\text{NOT TRIP}}), \text{fault location between P3 and P4} \quad (3.2)$$

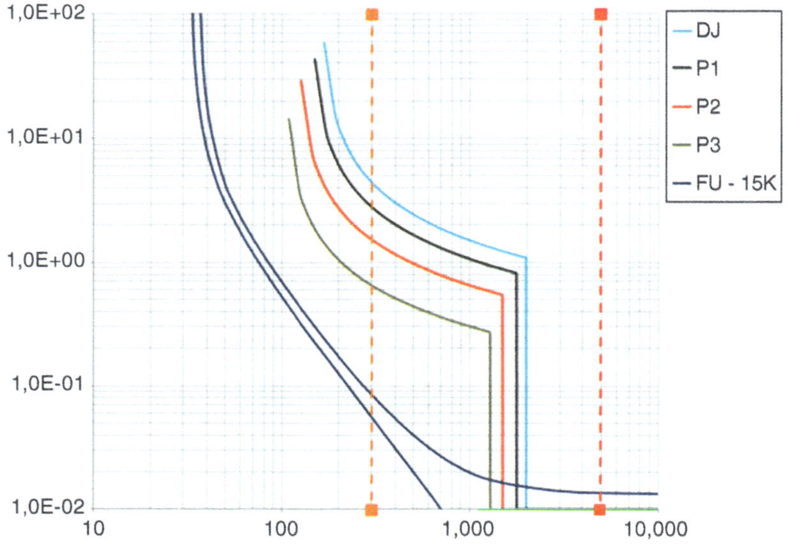

Fig. 3.22 Time coordination curves—Topology N° 7—selective system

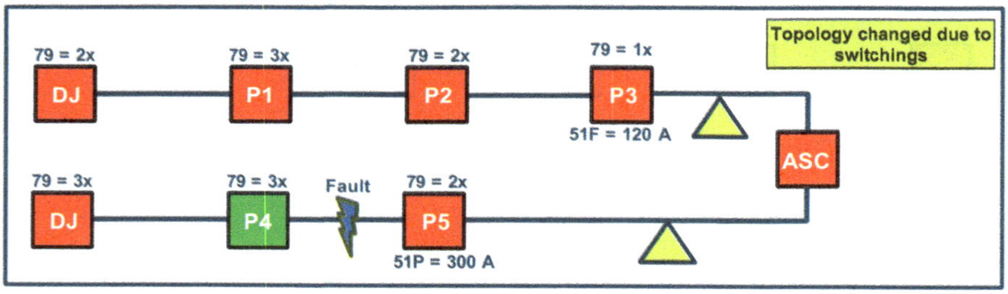

Fig. 3.23 Topology N° 7, system in the modified configuration

3.4 Self-Healing: Decentralized Operation

Decentralized systems of self-healing consist of using equipment in the distribution networks, as switches or reclosers, which, through their own information, perform actions independent of a central command or human interference. In this kind of operation, the equipment takes decisions according to its programming, after processing its own information, associated or not to data of the adjacent equipment, when there is communication between devices.

The use of remotely operated switches that remain opened in a steady-state normal configuration of the network is common in interconnections between different feeders or branches in a feeder. These devices provide agility for operators when

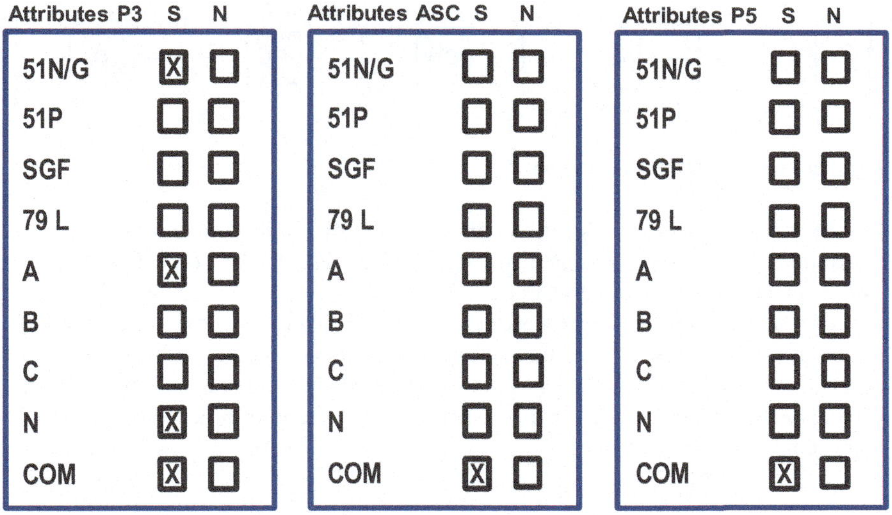

Fig. 3.24 Attributes for identification of the equipment signaled in SCADA—Topology N° 7

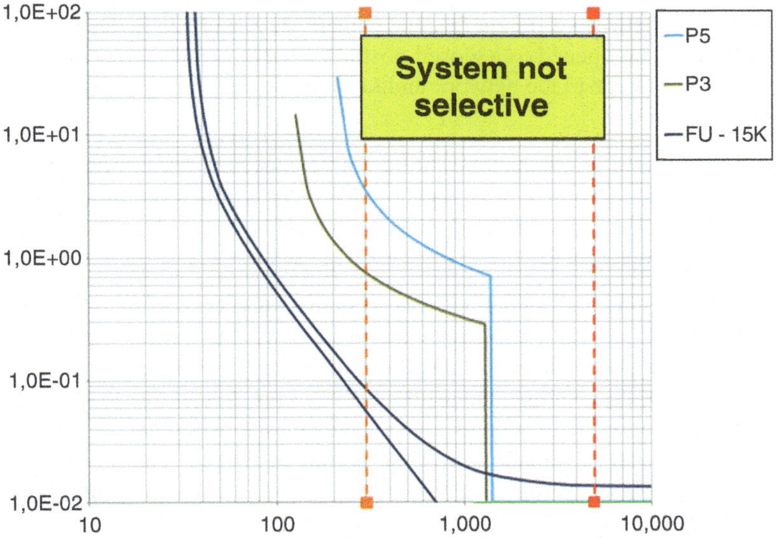

Fig. 3.25 Time coordination curves—Topology N° 7—uncoordinated system

there is an interruption of the electric energy supply or when a fault occurs, allowing part or all the system to be energized by a new source or a new electric path. The practice of installing remote control interconnection equipment, which we call the NO type (see Fig. 3.26), helps to reduce the continuity indicators, since they

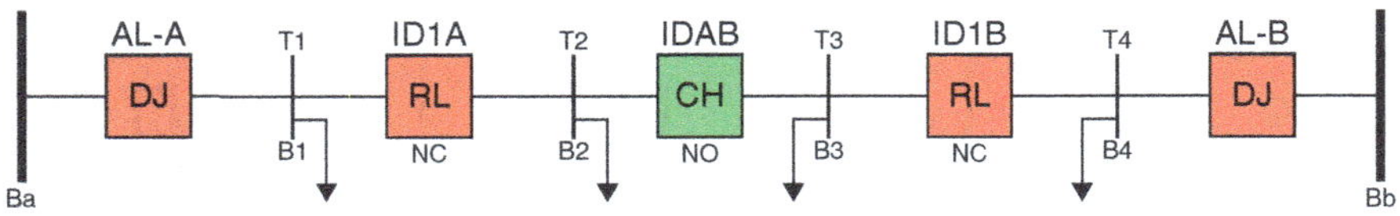

Fig. 3.26 Generic distribution network

provide multiple alternatives for reconfiguring the distribution networks. A generic distribution network is presented in Fig. 3.26, where the NO type IDAB device interconnects two distinct networks.

Logic and functions can be associated to the interconnection device IDAB, allowing, in the absence of voltage at one side of its connection points, the closing action, interconnecting the feeders and providing self-healing in a decentralized way. This decision-making process must respect certain criteria in order to energize only clear branches.

3.4.1 *Objective Functions and Constraints*

According to Fig. 3.26, a remotely controlled device placed at IDAB allows the interconnection between different feeders, named AL-A and AL-B. The switching of this device from an opened to a closed position needs to respect several criteria, in order to avoid damage to the network and keep the power quality. These criteria are described as follows:

- The thermal limit of conductors should not be exceeded for all branches.
- The magnitude of voltage at buses should be kept within a permissible range.
- A predetermined percentage of the overcurrent setting of protection devices should not be exceeded.

When performing the switching, the system that receives the loads must be capable to supply all the customers without inadmissible currents in conductors. This criterion must be attended during all day and every day of the year, in order that the logic of the equipment operates without constraints. In case the loading criterion is met, one must observe the second criterion which establishes that the magnitude of voltages in buses must remain within the permitted ranges. This is to ensure the quality of the power delivered to customers. In the same way as the loading criterion, the voltage criterion must be met for all load levels.

In order to ensure that all criteria are met, simulations must be performed to ensure correct operation. The objective function (FO) given by Eq. (3.3) maximizes the number of customers supplied, subject to constraints related to the limit of current in each branch, as defined in Eq. (3.4).

$$\mathrm{FO} = \max \mathrm{NCR} \tag{3.3}$$

where NCR is the number of restored consumers.

The object function is subjected to the constraints given in Eqs. (3.4) and (3.6). The constraint (3.4) is related to the limit of current in each branch:

$$I_{Tx} \leq I\mathrm{max}_{Tx} \times k_{Tx} \tag{3.4}$$

where:

$I\mathrm{max}_{Tx}$: current limit of cable T in branch x

k_{Tx}: permissible overload factor of branch x, usually between 0.85 and 1.30

I_{Tx}: current adjusted to the limit in branch x

The total current of the system, when interconnected, can be obtained according to Eq. (3.5):

$$I_{\mathrm{total}} = T1 \times I_{T1} + T2 \times I_{T2} + T3 \times I_{T3} + T4 \times I_{T4} \tag{3.5}$$

where

I_{total} is the total current in the period under analysis

I_{Tx}: current adjusted to the limit in branch x

Tx: digital status (0 or 1) recloser/switch interconnection branches

Regarding the permissible voltage at the buses, the constraint defined in Eq. (3.6) is applied for all buses:

$$V\mathrm{lim}_{\mathrm{min}} \leq V_{\mathrm{B}x} \tag{3.6}$$

where:

$V_{\mathrm{B}x}$: voltage at the bus x

$V\mathrm{lim}_{\mathrm{min}}$: minimum voltage allowed at bus x

3.4.2 Devices

The equipment adopted to perform the load transfer automatically in a decentralized system should have the following characteristics:

- It must be able to detect presence/absence of voltage at source and load sides.
- It must have current sensors with associated protection functions.
- It must be capable of processing control logics associated with the state of equipment and/or its protections.

Functional overview of the equipment adopted to perform the load transfer automatically is presented in Fig. 3.27.

The ability to process voltage information associated with overcurrent protection, as well as the possibility of developing control logic, allows the implementation of

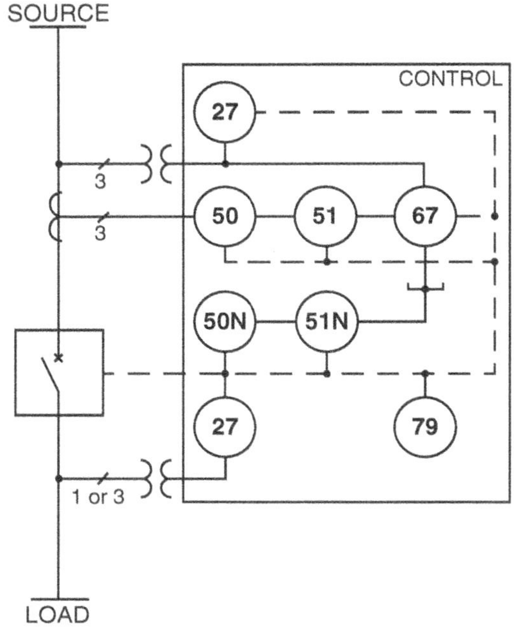

Fig. 3.27 Functional overview device [13]

the constraints presented in Eqs. (3.4–3.6) for the objective function in Eq. (3.3). Some protection functions can be used, such as:

- Reclosing relay, 79
- Phase time-overcurrent relay, 51P
- Phase instantaneous overcurrent, 50P
- Neutral time-overcurrent relay, 51N/G
- Neutral instantaneous overcurrent, 50N/G
- Sensitive ground fault or sensitive earth fault overcurrent, 51SGF or 51SEF
- Directional overcurrent relay, 67

Control and protection arrangements can be incorporated in order to meet the needs of the branches downstream the equipment. These functions do not necessarily trip on the equipment but indicate the logical condition of the arrangement that the equipment is supplying. This applies the correct logic to the current condition.

Commercial equipment, such as reclosers, have pre-programmed self-healing functions, allowing for quick setup with configured program logic, reducing factory acceptance tests or field acceptance tests and function test times. This function can be called Automatic Source Transfer.

The equipment in the distribution network may occasionally present independent logics to ensure that given constraints are not violated. For instance, when there is a fault in AL-A between DJ and ID1A (see Fig. 3.26), when closing IDAB an overload can occur. In order to ensure that the system does not violate this constraint, the undervoltage relay (27) on ID1A is used, ensuring that when there is an outage, the device performs the isolation of the branches.

As recommendation due the system operation mode, the equipment which has the self-healing function enabled must not have the automatic reclosing function enabled. This is because it is not possible to guarantee that the branch under fault is isolated, once the energization of the branch under fault could occur unduly. Furthermore, due to the possibility that the fault may not be the same as that identified by the original recloser, it is necessary to associate other protections previously cited, especially the protections related to voltage and current unbalance, with the purpose of ensuring safety and protection of equipment and population.

3.4.3 Systems with Communication

In order to increase the number of recomposed customers, one can improve the performance of the logic used by devices in the distribution network through the communication between them. A communication network between the devices provides the exchange of relevant information, such as the type of protection signalized, current and/or power measures in the instant before a fault, as well as the states of the devices. A typical arrangement of a communication system is presented in Fig. 3.28.

Through the exchange of information, it can be guaranteed the section under fault will be correctly isolated, then the distribution network can be energized by the new source through the normally opened (NO) switch. Leading manufacturers of protective equipment provide ways for real-time information exchange between identical devices through proprietary protocols. In addition, with the advent of new standards and protocols, information can be exchanged between different devices through

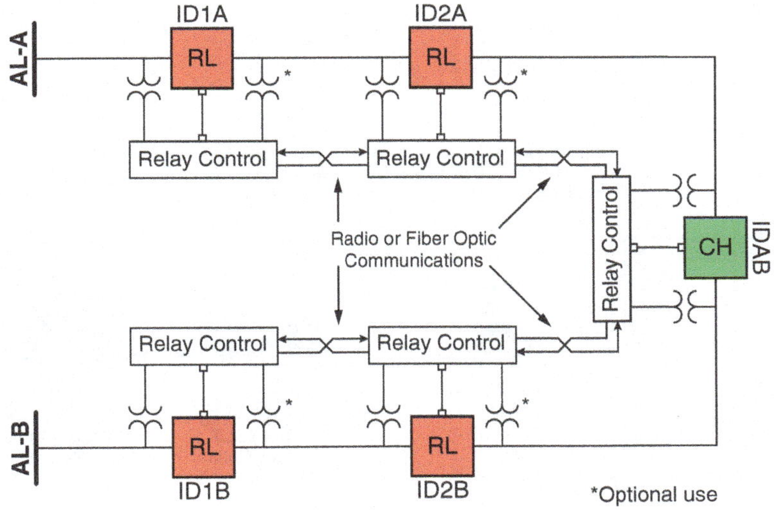

Fig. 3.28 Typical arrangement for communication system

standardized protocols, such as IEC103 and GOOSE, or proprietary protocols intended for communication between equipment.

The application of peer-to-peer communication provides gains in terms of performance in recomposing the system, as it provides agility and assertiveness in reconfiguration, ensuring correct isolation of the faulty branches. This is due to real-time information exchange between devices, which allows the identification of the type of protection tripped and the signaling of the state of the equipment included in the logical arrangement.

In order to ensure a safe operation in self-healing systems, it is recommended to integrate contingency logics when communication failure occurs. In this way, the condition of the fault is addressed according to necessity and priority, where the interconnecting device can operate in two different ways:

- Take no action during a communication failure, regardless of the distribution network conditions.
- Operate in a decentralized way in the operation mode without communication, taking controlled actions.

3.5 Self-Healing: Centralized Operation

The operation of a self-healing system through a centralized arrangement allows the use of different techniques when compared to the decentralized arrangement. Although the centralized philosophy is totally dependent on the existing communication system between the field equipment and the SCADA system, it allows the integration of the power supply restoration solution with several operating tools used in an operation center. The architecture of a centralized self-healing system [14] is presented in Fig. 3.29.

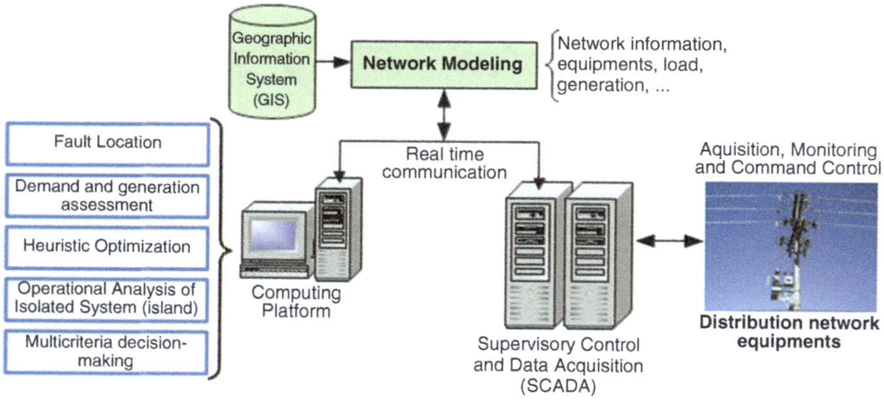

Fig. 3.29 General system architecture for automatic power restoration

The complexity of integrating the centralized system with the different computational platforms in the market should be highlighted. Many of the major international manufacturers currently offer fully integrated solutions, the so-called Advanced Distribution Management System (ADMS).

However, the centralized system allows the application of different algorithms running in the background, also being possible to have objective functions to minimize the impacts of the power interruption and maximize the reestablished customers.

3.5.1 Objective Functions and Constraints

Once the region of the system with fault has been identified, the possibilities of switching to isolate the fault and transfer loads are verified. In the case of load transfers, it is necessary to analyze which option is most appropriate when there is more than one feasible switching alternative. Thus, the definition of objective functions and constraints to evaluate which is the best option becomes necessary.

Several types of objective functions can be considered [15]. The most usual are those that consider the number of consumers and energy restored. Other factors can be considered, such as the continuity indicators and the energy losses resulting from the proposed configurations.

Among the constraints, the main issues are concerned with to avoid overloading the electrical elements, avoid violation of the protection settings, and keep the voltage levels within a prescribed range. Consideration should also be given to the limits of generation sources.

Based on these considerations, the following objective functions (OF1 and OF2) and constraints for the automatic restoration of electric power are defined:

First priority:

$$\mathrm{OF1} = \max \mathrm{NCR}^* \cdot w_1 + \mathrm{QER}^* \cdot w_2 \tag{3.7}$$

Second priority:

$$\begin{aligned}\mathrm{OF2} = \min\ & \mathrm{SAIFI_{exp}}^* \cdot w_3 + \mathrm{SAIDI_{exp}}^* \cdot w_4 + \mathrm{ENS_{exp}}^* \cdot w_5 \\ & + \mathrm{Losses_{exp}}^* \cdot w_6\end{aligned} \tag{3.8}$$

subject to:

- Maintaining the radiality of the feeders
- Feeders must not operate with overload
- The settings of protection equipment should not be violated
- The voltage limits should not be violated
- Distributed generators should operate according to their operational limits regarding active and reactive power
- The variables in Eqs. (3.7) and (3.8) represent:

- NCR: number of restored consumers
- QER: restored electrical energy (kWh)
- $SAIFI_{exp}$: frequency of expected interruption per consumer (failures/year), for the proposed configuration
- $SAIDI_{exp}$: equivalent duration of expected interruption per consumer (h/year), for the proposed configuration
- ENS_{exp}: expected energy not supplied per year (MWh/year), for the proposed configuration
- $Losses_{exp}$: expected energy losses (kWh), for the proposed configuration
- w_1, w_2, w_3, w_4, w_5, w_6: weights of the criteria
- *: indicates normalized values

The analysis of the objective functions is performed hierarchically. In cases where the alternatives have the same value for OF1 (Priority 1), OF2 (Priority 2) is applied like a tie-breaking criterion. In what follows, a description of how the values of each criterion are calculated is presented.

The number of consumers restored (NCR) is calculated by the sum of consumers who were interrupted and transferred to the other feeder, according to Eq. (3.9).

$$\mathrm{NCR} = \sum_{j=1}^{N_{\mathrm{tr}}} C_j \tag{3.9}$$

where:

- C_j: number of restored consumers in transformer j
- j: distribution transformer index
- N_{tr}: total number of restored distribution transformers

Similarly, the following equation is used to obtain the restored energy, in kWh:

$$\mathrm{QER} = \sum_{j=1}^{N_{\mathrm{tr}}} L_j \tag{3.10}$$

where L_j is the restored energy in transformer j.

The energy losses (Losses) and expected reliability indicators ($SAIFI_{exp}$, $SAIDI_{exp}$ and ENS_{exp}) for the proposed configuration are calculated according to the equations presented [16, 17]. The constraints are verified through the power flow solution of the proposed network configuration.

3.5.2 *Method of Multicriteria Decision-Making*

The objective of this stage of the optimization process is to determine the best alternative for energy restoration, when there is more than one option of load transfer or there is the possibility of islanded operation with distributed generation. The switchings that will be executed among the selected ones will be defined, according to the objective functions, without violating the constraints.

The proposal is to apply Eq. (3.7), choosing the solution that presents the highest value for OF1 (maximize function). In case of a draw, apply Eq. (3.8), choosing the solution that presents the lowest value for OF2 (minimize function). The Analytic Hierarchy Process (AHP) method is applied to solve this problem. For OF1, the following judgment matrix is defined:

$$M = \begin{array}{c} \\ C_1 \\ C_2 \end{array} \begin{array}{c} \begin{array}{cc} C_1 & C_2 \end{array} \\ \begin{bmatrix} 1 & 3 \\ 1/3 & 1 \end{bmatrix} \end{array} \quad (3.11)$$

where:

- $C1$: NCR criterion
- $C2$: QER criterion

Through the application of the proposed method, the following weight matrix can be obtained:

$$w = \begin{array}{c} w_1 \\ w_2 \end{array} \begin{bmatrix} 0.68 \\ 0.32 \end{bmatrix} \quad (3.12)$$

For the OF2 function , it was defined:

$$M = \begin{array}{c} \\ C_3 \\ C_4 \\ C_5 \\ C_6 \end{array} \begin{array}{c} \begin{array}{cccc} C_3 & C_4 & C_5 & C_6 \end{array} \\ \begin{bmatrix} 1 & 3 & 5 & 7 \\ 1/3 & 1 & 3 & 5 \\ 1/5 & 1/3 & 1 & 3 \\ 1/7 & 1/5 & 1/3 & 1 \end{bmatrix} \end{array} \quad (3.13)$$

where:

- C_3: $\text{SAIFI}_{\text{exp}}$ criterion
- C_4: $\text{SAIDI}_{\text{exp}}$ criterion
- C_5: ENS_{exp} criterion
- C_6: $\text{LOSSES}_{\text{exp}}$ criterion

For this case, then, the following weight matrix can be obtained:

$$w = \begin{array}{c} w_3 \\ w_4 \\ w_5 \\ w_6 \end{array} \begin{bmatrix} 0.56 \\ 0.26 \\ 0.12 \\ 0.06 \end{bmatrix} \quad (3.14)$$

3.6 Practical Applications

In this section practical applications are described, considering decentralized and centralized systems as examples.

3.6.1 Decentralized System—Logic of Energization via NO Without Communication

The basic way to connect the NO equipment consists of a simple logic of verification if there is or not voltage at the source or the load sides. After a predefined time, the energization is performed. Figure 3.30 presents a diagram of a logic with opening command, where the variables represent:

- 27S—trip of the undervoltage relay at the load side
- 27L—trip of the undervoltage relay at the source side
- POS_OPEN—indication of the open position of the device
- CMD_OPEN—point to enter in the device opening logic

The timer *T*1 must be set so that its time is longer than the longer reclosing time of the devices that are part of the network under switching. For instance, consider the network presented in Fig. 3.31, the maximum reclosing times are:

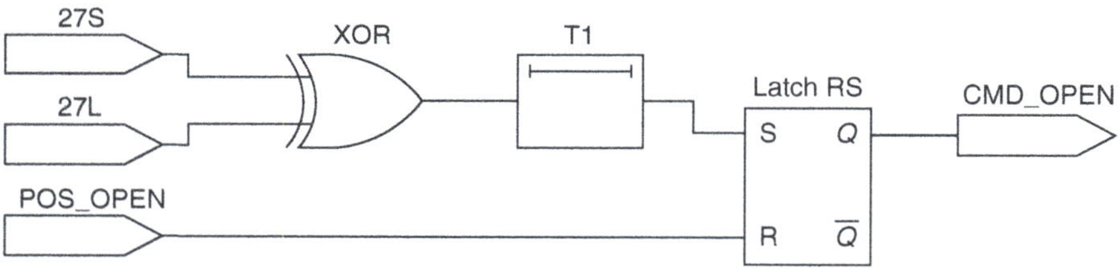

Fig. 3.30 Logic with opening command

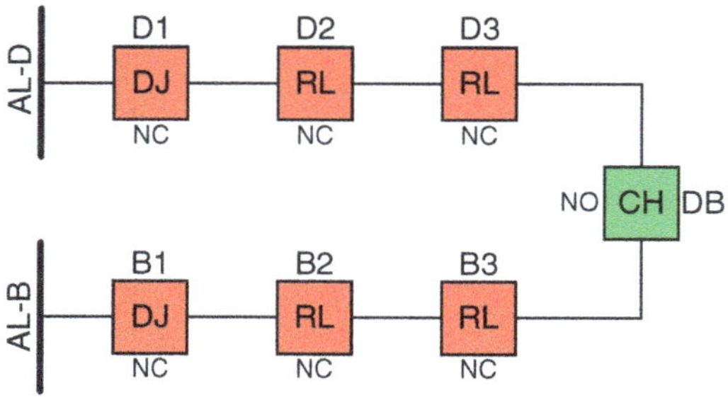

Fig. 3.31 Network example for the application of time delay

- $t\max_{D1} = 130$ s
- $t\max_{D2} = 65$ s
- $t\max_{D3} = 30$ s
- $t\max_{B1} = 150$ s
- $t\max_{B2} = 90$ s
- $t\max_{B3} = 45$ s

In this way, the time considered in *T*1 must be greater than 150 s, which is the longer reclosing time (device B1). As a coordination criterion, 15 s can be added to this time in order to guarantee possible failures, totalizing for this case 165 s, which is set in T1. This ensures that the closing of the DB switch will occur only after all possible reclosing attempts have been performed.

3.6.2 Centralized System

Figure 3.32 illustrates a hypothetical distribution network to demonstrate the energy restoration methodology, defining the switchings upstream and downstream of the fault, and including the possibility of post-contingency island operation, as there is a distributed generation in AL2.

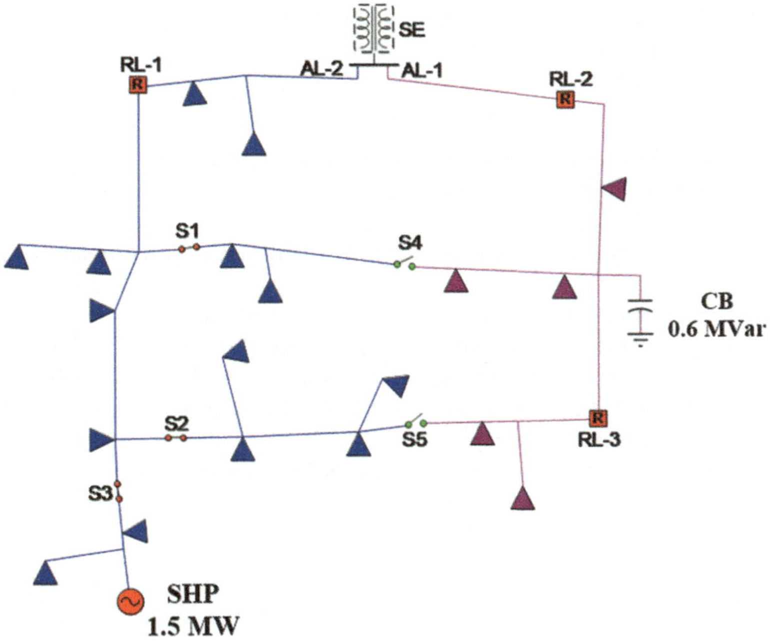

Fig. 3.32 Distribution feeders

The system has two feeders, where reclosers (R) and switches (S) normally closed (NC), in red, and switches NO, in green, are remotely controlled. It also has a small hydropower plant (SHP) representing the distributed generation source.

In order to illustrate the application of the methodology, three case studies are presented: (a) fault in the downstream branch of RL-1, without the possibility of islanded operation; (b) fault in the downstream branch of RL-1, with the possibility of islanded operation; and (c) fault between the DJ of the AL-1 and RL-1, with the possibility of islanded operation.

In all cases, the following objective function and constraint values were considered:

Priority 1

$$\mathrm{FO1} = \max \mathrm{NCR} \times 0.68 + \mathrm{QER} \times 0.32 \tag{3.15}$$

Priority 2

$$\mathrm{FO2} = \min \mathrm{SAIFI}_{\mathrm{exp}} \times 0.56 + \mathrm{SAIDI}_{\mathrm{exp}} \times 0.26 + \mathrm{ENS}_{\mathrm{exp}} \times 0.12 + \mathrm{LOSSES}_{\mathrm{exp}} \times 0.06 \tag{3.16}$$

subject to:

- Maintaining the radiality of the feeders
- Loading allowed in the network and equipment $\leq$100%
- Loading in relation to the settings of overcurrent protection devices $\leq$90%
- Voltage drop allowed in the medium voltage network $\leq$7%
- Capacity of distributed generation $\leq$100%

3.6.2.1 Fault in the Downstream Branch of RL-1, Without the Possibility of Islanded Operation

Figure 3.33 illustrates the system considering a fault in the downstream branch of RL-1.

The results obtained for this fault are summarized in Table 3.1.

In this case, there is only one switching option for each possibility of load transfer, totaling two viable switching to be executed. It should be noted that the possibility of islanded operation was not considered, causing a greater number of consumers without energy supply.

Figure 3.34 shows the switches indicated for load transfers.

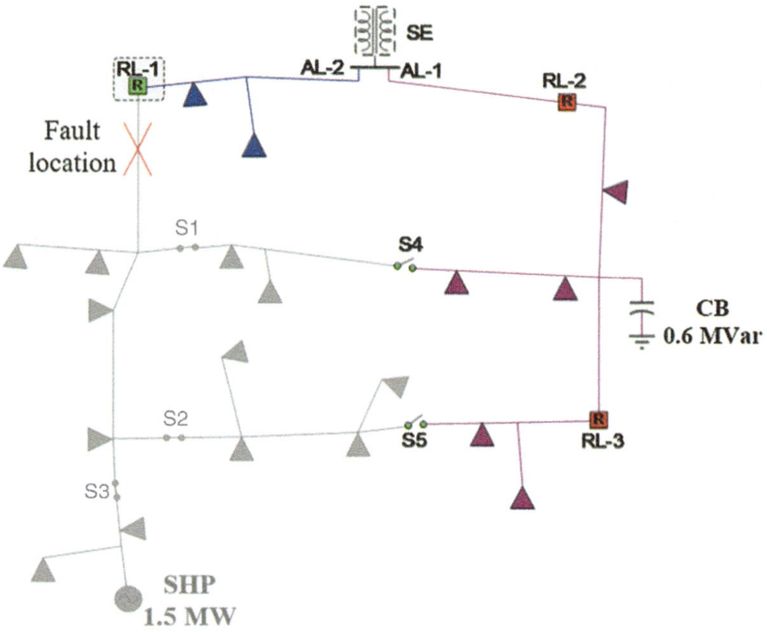

Fig. 3.33 System with a fault in the downstream branch of RL-1

3.6.2.2 Fault in the Downstream Branch of RL-1, with the Possibility of Islanded Operation

In this case study the possibility of islanded operation was considered, serving part of the consumers with the SHP. Table 3.2 summarizes the results obtained for this condition, considering the distribution network shown in Fig. 3.33.

From the results, one can verify that the energy supply was restored for the customers downstream of switch S3 through the SHP, being this an option for the cases in which the post-contingency island mode is the only way to restore the energy while repairing the fault. Such an option still lacks more elaborate studies to become applicable in practice.

Figure 3.35 illustrates the switches indicated for load transfers and islanded operation carried out by the SHP.

3.6.2.3 Fault Between the DJ of AL-1 and RL-1, with the Possibility of Islanded Operation

Figure 3.36 illustrates the same test network, however with a fault at a different location, in order to demonstrate the methodology in the case where there is more than one load transfer option from the same normally closed switch.

Table 3.1 Results of switchings upstream and downstream of the fault—case (a)

Contingency conditions					
Acting device RL 1 Disconnection of the SHP Customers without energy supply: 1200 Interrupted energy: 2486 kWh					
Switchings upstream of the fault					
Device already actuated					
Switchings downstream of the fault					
Open S 1 Close S 4 Open S 2 Close S 5					
Load transfers analysis					
Device reference	Switching	New configuration indicators	Value	OF	Constraints
S 1	Open S 1 Close S 4	NCR	200 cust.	Do not apply	No violation
		QER	301 kWh		
		$SAIFI_{exp}$	11 times		
		$SAIDI_{exp}$	15 h		
		ENS_{exp}	79 MWh		
		$Losses_{exp}$	322 kWh		
		Loading	77%		
		Protection	67%		
		Voltage drop	4%		
S 2	Open S 2 Close S 5	NCR	400 cust.	Do not apply	No violation
		QER	1211 kWh		
		$SAIFI_{exp}$	10 times		
		$SAIDI_{exp}$	14 h		
		ENS_{exp}	83 MWh		
		$Losses_{exp}$	427 kWh		
		Loading	68%		
		Protection	78%		
		Voltage drop	5%		
Condition after switchings					
Customers without energy supply: 600 Interrupted energy: 974 kWh					

The results for this case study are presented in Table 3.3.

The first and second switching results, which are open RL1 and close S4, and open RL1 and close S5, are both feasible. However, the third switching, which is open RL1 and operates the SHP in the islanded mode, is not possible, since it violates the nominal generation capacity in 65%.

Since the two feasible options to meet the same load present the same value for the first objective function, OF1 (NCR and QER), the second objective function OF2 ($SAIFI_{exp}$, $SAIDI_{exp}$, ENS_{exp} and $Losses_{exp}$ for the new configuration) is evaluated

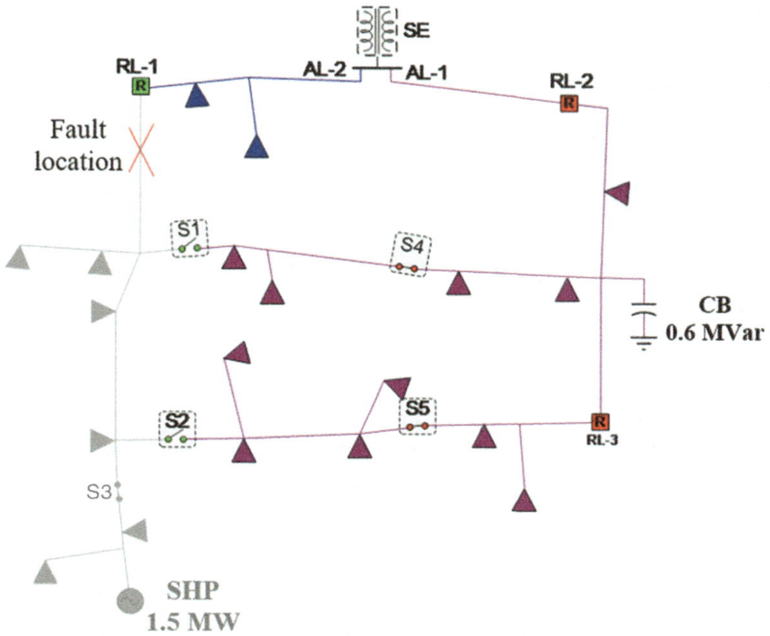

Fig. 3.34 Indication of the load transfer switchings–case (a)

as a tie-breaking criterion. The resulting best alternative, which presented a lower value for OF2, is the following switching action: open RL1 and close S4.

Figure 3.37 illustrates the switches indicated for load transfers in this case study, considering the best solution found.

3.7 Conclusion

This chapter presented important topics regarding the automatic energy restoration in distribution network, from which one can verify that there may be different arrangements for the implementation of a self-healing system. Details such as the operating philosophy adopted by power utility and the operational procedures are very important to define the best operating strategy for this kind of system.

The approaches involving centralized and decentralized systems have advantages and disadvantages. Regarding centralized system, it allows a greater integration of data and solutions with tools already present in Operation Centers. On the other hand, the decentralized system has a greater autonomy regarding communication failures and faster performance to restore the energy supply.

Finally, as one can see, the operation of a self-healing system is quite complex and needs to be carefully planned and integrated among the most diverse areas in an electric power utility.

Table 3.2 Results of switchings upstream and downstream of the fault—case (b)

Contingency conditions					
Acting device RL 1 Disconnection of the SHP Customers without energy supply: 1200 Interrupted energy: 2486 kWh					
Switchings upstream of the fault					
Device already actuated					
Switchings downstream of the fault					
Open S 1 Close S 4 Open S 2 Close S 5 Islanded operation SHP					
Load transfers analysis					
Device reference	Switching	New configuration indicators	Value	OF	Constraints
S 1	Open S 1 Close S 4	NCR	200 cust.	Do not apply	No violation
		QER	301 kWh		
		$SAIFI_{esp}$	11 times		
		$SAIDI_{esp}$	15 h		
		ENS_{esp}	79 MWh		
		$Losses_{exp}$	322 kWh		
		Loading	77%		
		Protection	67%		
		Voltage drop	4%		
S 2	Open S 2 Close S 5	NCR	400 cust.	Do not apply	No violation
		QER	1211 kWh		
		$SAIFI_{esp}$	10 times		
		$SAIDI_{esp}$	14 h		
		ENS_{esp}	83 MWh		
		$Losses_{exp}$	427 kWh		
		Loading	68%		
		Protection	78%		
		Voltage drop	5%		
S 3	Open S 3 Islanded operation SHP	NCR	200 cust.	Do not apply	No violation
		QER	450 kWh		
		FEC_{esp}	11 times		
		DEC_{esp}	16 h		
		ENS_{esp}	88 MWh		
		$Losses_{exp}$	298 kWh		
		SHP capacity	30%		
		Loading	40%		
		Protection	40%		
		Voltage drop	1%		
Condition after switchings					
Customers without energy supply: 400 Interrupted energy: 524 kWh					

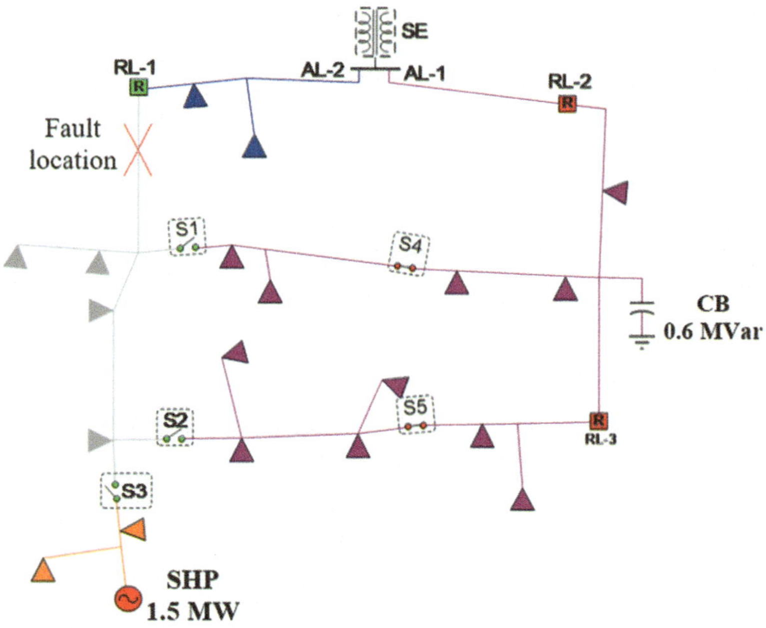

Fig. 3.35 Indication of the load transfer switchings—case (b)

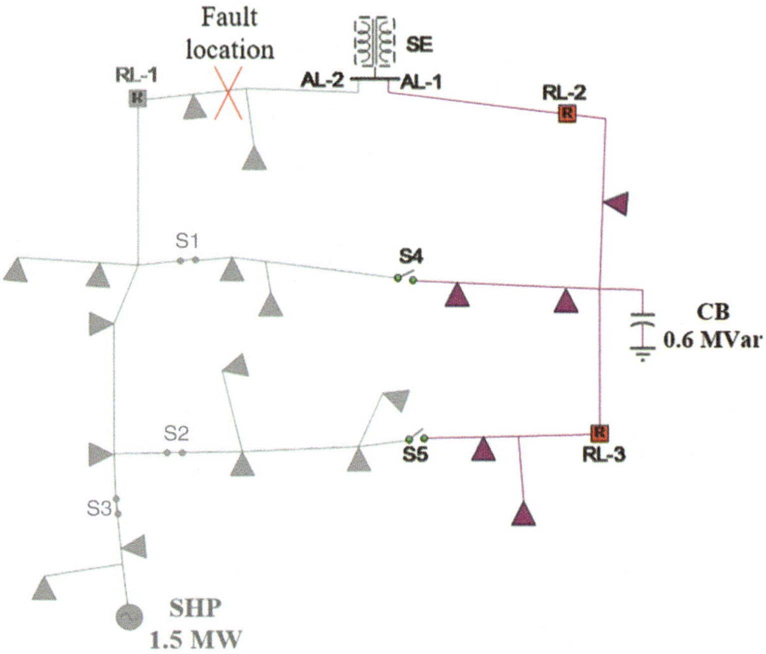

Fig. 3.36 Distribution system with a fault in the feeder AL-2

Table 3.3 Results of switchings upstream and downstream of the fault—case (c)

Contingency conditions					
Acting device DJ AL 1 Disconnection of the SHP Customers without energy supply: 1400 Interrupted energy: 2686 kWh					
Switchings upstream of the fault					
Device already actuated					
Switchings downstream of the fault					
Turn off RL 1 Close S 4					
Load transfers analysis					
Device reference	Switchings	New configuration indicators	Value	OF	Constraints
RL 1	Open RL 1 Close S 4	NCR	1200 cust.	OF 1 = 1.0 OF 2 = 0.90	No violation
		QER	2486 kWh		
		$SAIFI_{exp}$	16 times		
		$SAIDI_{exp}$	24 h		
		ENS_{exp}	159 MWh		
		$Losses_{exp}$	500 kWh		
		Loading	95%		
		Protection	85%		
		Voltage drop	6%		
	Open RL 1 Close S 5	NCR	1200 cust.	OF 1 = 1.0 OF 2 = 0.99	No violation
		QER	2486 kWh		
		$SAIFI_{esp}$	18 times		
		$SAIDI_{esp}$	26 h		
		ENS_{esp}	158 MWh		
		$Losses_{exp}$	637 kWh		
		Loading	98%		
		Protection	88%		
		Voltage drop	7%		
	Open RL 1 SHP islanded operation	NCR	1200 cust.	Do not apply	Violation in the SHP capacity
		QER	2486 kWh		
		$SAIFI_{exp}$	13 times		
		$SAIDI_{exp}$	19 h		
		ENS_{exp}	76 MWh		
		$Losses_{exp}$	319 kWh		
		SHP capacity	165%		
		Loading	95%		
		Protection	85%		
		Voltage drop	6%		
Condition after switchings					
Customers without energy supply: 200 Interrupted energy: 200 kWh					

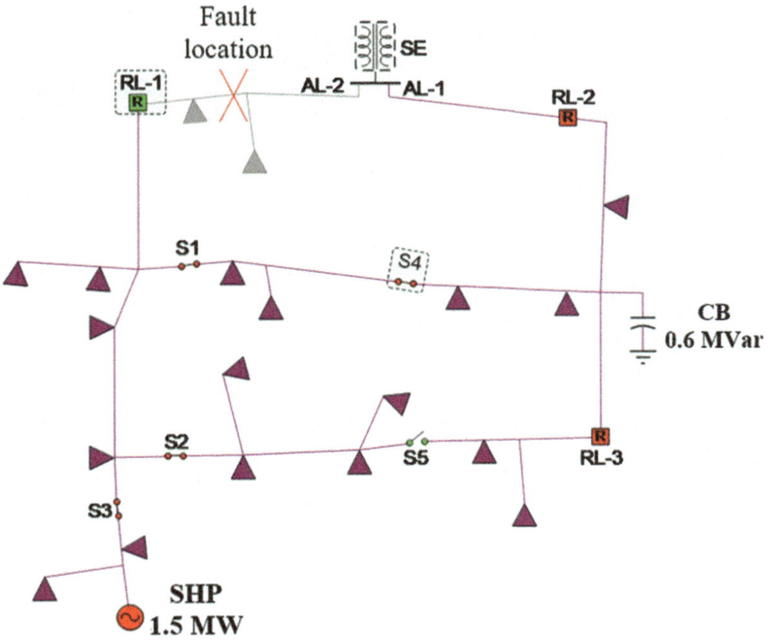

Fig. 3.37 Indication of the load transfer switchings—case (c)

References

1. Kagan N, Oliveira CCB (1996) Reconfiguração de sistemas de distribuição de energia elétrica através de ferramenta para solução de problemas de decisão com múltiplos objetivos e incertezas. In: CBA—Congresso Brasileiro de Automática. Anais, São Paulo
2. Zidan A, El-Saadany E (2012) A cooperative multiagent framework for self-healing mechanisms in distribution systems. IEEE Trans Smart Grid 3(3):1525–1539
3. Arefifar S, Mohamed Y-R, El-Fouly T (2013) Comprehensive operational planning framework for self-healing control actions in smart distribution grids. IEEE Trans Power Syst 28 (4):4192–4200
4. Wang Z, Wang J (2015) Self-healing resilient distribution systems based on Sectionalization into microgrids. IEEE Trans Power Syst 30(6):3139–3149
5. Garcia VJ (2005) Algoritmos para otimização multiobjetivo aplicados ao problema de restauração do serviço em redes de distribuição de energia elétrica. Dissertation (Doctorate in Electrical Engineering) Universidade Estadual de Campinas, Campinas
6. Sperandio M (2008) Planejamento da Automação de Sistemas de Manobra em Redes de Distribuição. Dissertation (Doctorate in Electrical Engineering)—Universidade Federal de Santa Catarina, Florianópolis
7. Brown RE (2008) Impact of smart grid on distribution system design. In: IEEE power and energy society general meeting—conversion and delivery of electrical energy in the 21st century. Anais. IEEE PES
8. Falcão DM (2010) Integração de Tecnologias para Viabilização da Smart Grid. In: III Simpósio Brasileiro de Sistemas Elétricos (SBSE). Anais Belém

9. Bernardon DP (2007) Novos métodos para reconfiguração das redes de distribuição a partir de algoritmos de tomadas de decisão multicriteriais. Dissertation (Doctorate in Electrical Engineering)—Universidade Federal de Santa Maria, Santa Maria
10. Ferreira PM, et al. (2013) Ilhamento Voluntário de PCH para Melhorar os Indicadores de Qualidade DEC e FEC no Sistema Elétrico de Sua Região de Influência. In: VIII CIERTEC. Anais... Fortaleza
11. Dewadasa M, Ghosh A, Ledwich G (2011) Islanded operation and system restoration with converter interfaced distributed generation. In: Innovative Smart Grid Technologies Asia (ISGT) Anais IEEE PES
12. Li D, Wang S, Zhan J, Zhao Y (2011) A self-healing reconfiguration technique for smart distribution networks with DGs. In: International Conference on Electrical and Control Engineering (ICECE) Anais Yichang
13. Schweitzer Engineering Laboratories, Advanced recloser control manual of SEL-651-R
14. Kongoli F (2012) Automation. In Tech, Rijeka. ISBN: 978-953-51-0685-2
15. Popov DS, Ciri RM (1999) A multi-objective algorithm for distribution networks restoration. IEEE Trans Power Delivery 14(3)
16. Willis HL (2004) Power distribution planning reference handbook, 2nd edn. CRC Press, Boca Raton
17. Bernardon DP, Pfitscher LL, Canha LN, Mello APC, Abaide AR, Sperandio M, Garcia VJ, Ramos MJS (2015) Sistemas de Distribuição no Contexto das Redes Elétricas Inteligentes. AGEPOC, Santa Maria

Chapter 4
Volt/Var Control

A. P. C. Mello, G. S. da Silva, L. N. Canha, and Daniel Pinheiro Bernardon

4.1 Introduction

The modernization of the electric power system with Smart Grids, especially in the energy distribution sector, is being characterized by the increase and integration of automated equipment, two-way communication infrastructure and monitoring, processing and control systems, which can be managed by an Advanced Distribution Management System (ADMS).

The adhesion of these systems with the use of intelligent sensors, communication and control allows the state of the electrical network to be fully observed and promotes the development of new solution techniques for numerous decision support applications. One of these applications is the voltage and reactive power control (volt/var control), which plays an essential role in the electrical power system operation.

The volt/var control enables to maintain the voltage in suitable levels at all distribution feeder points, considering the most diverse system operating conditions [1, 2]. In this sense, the conventional equipment, such as the transformer with on-load tap-changer (OLTC) and the capacitor banks, has been the main devices used to correct the voltage violations in the distribution networks [3].

The increasing interest in the volt/var control in distribution systems is also due to the wide use of the Distributed Energy Resources (DER) from renewable sources, such as photovoltaic generation, which can change the energy quality of the

---

A. P. C. Mello (✉) · G. S. da Silva
Federal University of Pampa, Alegrete, Rio Grande do Sul, Brazil
e-mail: anamello@unipampa.edu.br; guilhermesiva@unipampa.edu.br

L. N. Canha · D. P. Bernardon
Technology Center, Federal University of Santa Maria, Santa Maria, Rio Grande do Sul, Brazil
e-mail: lucianecanha@ufsm.br; dpbernardon@ufsm.br

D. P. Bernardon, V. J. Garcia (eds.), *Smart Operation for Power Distribution Systems*, https://doi.org/10.1007/978-3-319-93922-3_4

customer, especially on the voltage levels. Similarly, the advent of new technologies such as the plug-in electric vehicles, which need to be connected to the grid from time to time to recharge the battery, can increase the energy demand and, consequently, can cause overloads and inadequate voltages on the distribution network.

With the imminent rise of new technologies, the volt/var control may also be a solution to voltage control applications with the distributed generation and electric vehicles, since these systems can also inject or absorb reactive power necessary for the voltage maintenance through the inverter, which connect the generation source with the distribution network [4]. The use of the inverter can contribute with the voltage regulation when the active power injected by the renewable sources reaches 15% of the nominal DG power or more and becomes essential when the active power reach 30% of the nominal DG power [5].

Some essential requirements to perform the equipment adjustments, considering a real-time volt/var control implementation, are presented in [6, 7], which are:

- DMS central control system with SCADA system support for the acquisition and processing of real-time measurements from the field controllable device
- Implementation of the three-phase unbalanced power flow for the distribution network validation and optimization
- Recent data record of the advanced communication infrastructure between the field controllable device (voltage regulators, capacitor banks, distributed generators, as well as other energy resources)
- Remotely controllable devices
- Efficient and modern communication system

One of the most important responsibilities of the grid utility is to keep the final consumer voltage within suitable limits [3]. Some situations, such as the fast growth of electricity demand, the connection/disconnection of the distributed generators, the electric vehicles, the network topology modification with the reconfiguration, and self-healing, require fast control performance to preserve the voltage levels in appropriate values. In addition, the voltage control also becomes extremely complex in large feeders with seasonality electric loads (summer and winter) as, for example, on rural feeders.

In this context, automated and coordinated volt-var control structures are a way to improve the distribution system operation, which can be managed in the distribution operation centers or in the substations. These architectures aim to be more effective, where the actions should has a systemic impact on the voltage levels of the distribution network.

4.2 Voltage and Reactive Power Control (Volt/Var Control)

4.2.1 Basic Concepts

The volt/var control study began in the late 1970s, with the search for new solutions due to increased energy and the global energy crisis. Traditionally, devices used on the voltage and reactive power control, such as voltage regulators, transformers with OLTC and capacitor banks, were individually and statically controlled, where the control between the devices are not coordinated. These devices were used local controls with fixed adjustments that they could not follow the changes of the daily load profiles.

With the control devices evolution from the 1990s, new approaches were introduced to the volt-var problem, including communication and monitoring of existing devices in distribution networks via Supervisory Control and Data Acquisition (SCADA). This evolution allowed a better use of the devices functions, giving rise to a centralized control, but that continued in an uncoordinated way, with individual actions that does not communicate with other equipment.

The main advantages of this approach in comparison to the fixed adjustment controllers are the more precise adjustment of the equipment and the possibility to disconnect the equipment when there are changes in the network or emergency situations. However, besides the centralized control require an efficient communication infrastructure, they have a limited capacity to adapt to dynamic changes in network operation, due to the system being based on previously defined rules. At the same time, an action performed in a device can be contrary regarding the actions of other equipment, and it is difficult to determine the adjustments in equipment that optimize the operation of the distribution network [7].

Over the last two decades, the electrical networks modernization has been intensified with the deployment of new smart technologies in the electrical networks, such as Distribution Automation Systems (DA), Distribution Management System (DMS), and of Advanced Metering Infrastructure (AMI), which encouraged the development of new strategies for volt/var control.

The DA concept refers to the use of automated control techniques in an intelligent, fully controlled and flexible distribution system that can help to a more efficient and reliable network operation. These characteristics are obtained because the DA system represents the integration of substation automation, feeder automation and consumer automation [8], as shown in Fig. 4.1.

The DMS is responsible to interconnect the advanced network automation from monitoring and control applications using an intelligent computing DMS. This DMS can directly operate the field equipment controllers or show the execution options to the operator, aiming to guarantee the system operation with security, integrity, and efficiency.

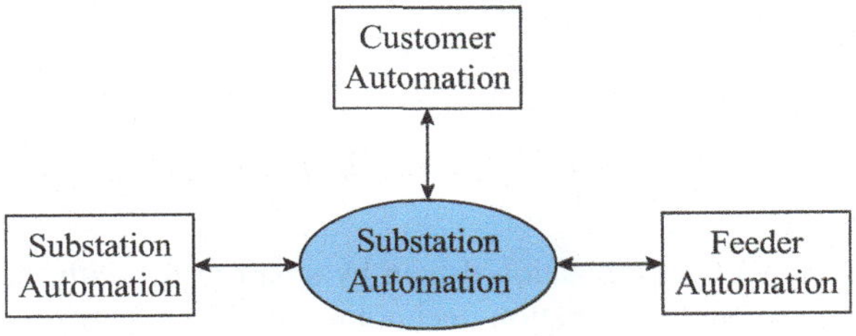

Fig. 4.1 Distribution automation states [8]

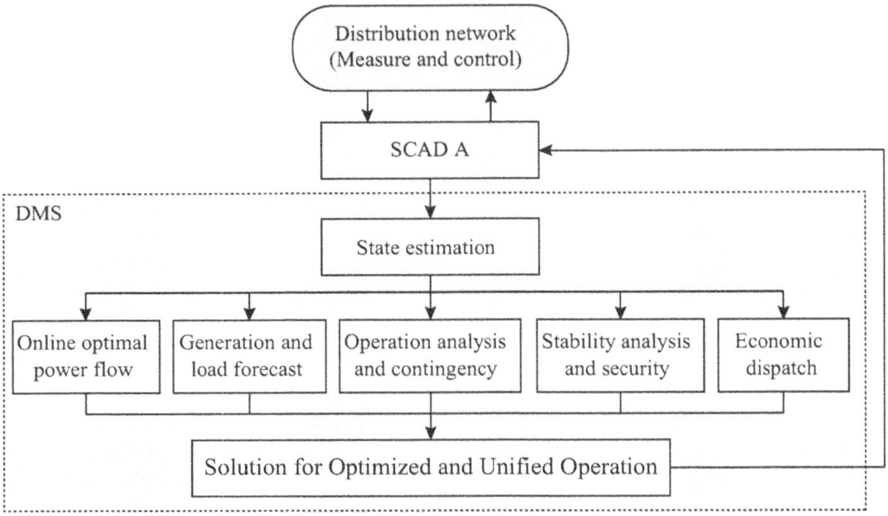

Fig. 4.2 Structure of the SCADA-DMS system for online control solution

Nevertheless, a central DMS management system is required for a unified system-wide management and operation, including operation of the volt/var control, and to coordinate the operations of intelligent and controllable equipment throughout the network. Figure 4.2 shows some specific functions of a DMS system. Field equipment transmits voltage and current data information through the Remote Terminal Unit (UTR) with SCADA communication channel to the DMS system, which centralizes operational decisions.

The control process occurs as follows: initially, the voltage and current data of all the system nodes are acquired as well as the updated operating states of all the control devices, which are transmitted via SCADA communication. These information are updated in the distribution network model of the main management systems, such as DMS. The DMS identifies the network problems such as voltage violations, excessive conductor overloads, equipment overuse/underuse, and evaluates potential corrections in decision-making to achieve a given optimization goal.

When the optimization objective is achieved, the DMS sends the required modifications to the field controllable equipment (transformers, regulators, capacitor banks, and generators, among others) to the SCADA, which take on new adjustment positions in the system.

4.2.2 *Volt/Var Control Objectives and Challenges*

The VVC optimization driven by smart grids, with the advances of the communication and automation technologies of the network equipment, can allow the implementation of new challenges, such as increased connection of small distributed generation; energy storage; electric vehicles connection; and smarter operation and control of the network equipment.

Among the main objectives of the VVC are [9, 10]:

- Keep the voltages at suitable levels
- Reduce the demand or allow the conservation voltage reduction to the load management
- Dynamic voltage control with adhesion of distributed generation from renewable sources and/or energy storage systems
- Voltage maintenance after reconfiguration and/or self-healing

4.2.2.1 Conservation Voltage Reduction

In the face of the smart grids emergence, many efficiency-enhancement processes of the distribution systems operation have stood out. One of these processes is the voltage reduction as a way of load managing (CVR, Conservation Voltage Reduction). The CVR refers to the reduction of the voltage supplied in the distribution networks to the lowest possible level within the appropriate voltage range to reduce the peak consumption of the feeders and to improve the energy savings for the distributor at times of high consumption.

The CVR factor calculation can be obtained by applying Eq. (4.1), which indicates the energy savings from the percentage difference between the demand reduction and voltage corresponding to the original system, without CVR application and the model that represents the operation with CVR.

$$\mathrm{CVR} = \frac{\%\mathrm{Energy\ \ Reduction}}{\%\mathrm{Voltage\ \ Reduction\ \ Difference}} \tag{4.1}$$

However, through a voltage reduction to the adequate value of the lower limit, voltage violations can occur in minor load variations, which can result in damage to equipment connected to the electrical distribution network.

4.2.3 Distributed Generation from Renewable Sources

Distributed Generation (DG) can be understood as the electricity production directly connected to the distribution network or to the local consumers' network [11]. The DG insertion in the electrical distribution network modifies the power flow originally in a passive system to an active system, representing a bidirectional power flow, that can directly affect the voltage regulation equipment. In addition, the DG size, the connection point, the system impedance, and the inverter operating mode are also factors that determine the DG impact on the system voltage regulation.

The DG connection from intermittent renewable sources in the feeders, such as wind and solar, can increase the overvoltage condition in the distribution system at the DG connection point, or also contribute to undervoltage, due to the DG disconnection or the shading in the case of photovoltaic generation, interrupting the energy generation. As power generation from renewable sources is not predictable, voltage and reactive power need to be controlled more frequently and accurately to match the output profile of renewable energy sources.

In addition to the critical overvoltage problem, given a significant DG share in the network, undesired operation of voltage regulators with line compensation can also occur due to reverse flow and changes in measured line current values [12].

4.2.3.1 Energy Storage Systems

A growing trend in distribution systems with the participation of distributed generation from renewable sources is the use of energy storage systems for the energy surplus of the distributed generators. In cases of large-scale stored energy, it is possible to rapidly inject energy into the system at times of greater demand, and replace or avoid the construction of new power plants for this purpose.

The injection of active power through energy storage systems, usually associated with batteries, can cause overvoltages during periods of low energy consumption. These systems can cause conflicts between traditional VVC actions, causing undesirable switching of control equipment. This way, prospecting future regulation and business models and also modifying the operation way of the distribution networks equipment, one can emulate the Volt/VAr/Watt control as an alternative to define the use of the storage systems depending on the control need of the distribution network [13].

4.2.3.2 Support for Reconfiguration and Self-Healing Operation

The reconfiguration of distribution networks consists of changing the network topology with the switches maneuvers, allowing the transfer of load from one feeder to the other in order to achieve the best possible efficiency in the network operation, with greater reliability and safety.

The VVC can also provide flexibility in reconfiguration and self-healing of distribution network operations. For example, in a permanent feeder fault event, changes in the network topology will occur with the change of the switches states for insulation of the fault area. These changes will modify the power flow, and the VVC will necessarily modify the field equipment settings to maintain proper voltage levels.

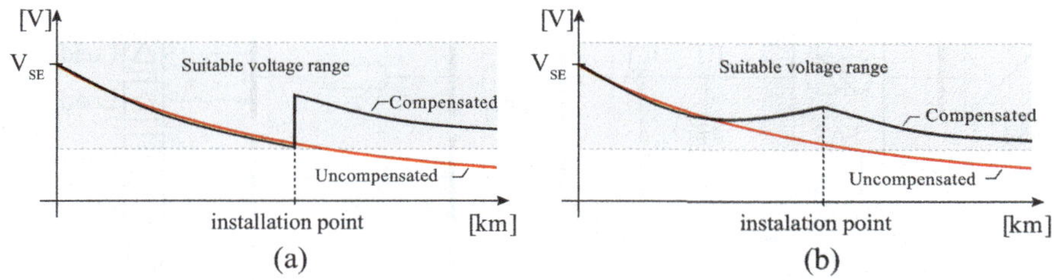

Fig. 4.3 Impact representation of the connected equipment in the distribution network. (**a**) Series-connected equipment. (**b**) Shunt-connected equipment

4.3 Equipment for Volt/Var Control

Several equipment has been developed and improved to safely and efficiently meet the power availability of distribution networks and maintain voltage within acceptable limits. Among the conventional regulating devices that change the voltage according to the load profile are: transformers with automatic switching under load, distribution transformers with derivations; capacitor banks; and voltage regulators.

These control devices can be classified into two main categories: serial compensation devices and shunt compensation devices. Serial compensation devices have the ability to correct the voltages downstream of their installation, as shown in Fig. 4.3a. The series compensation devices traditionally used are the voltage regulators; however, there are also series capacitors and series reactive static compensators, among others.

Devices with shunt compensation have the ability to correct the voltages upstream and downstream of their installation, as shown in Fig. 4.3b. The traditional shunt compensation devices used are capacitor banks, but there are also distributed generation, shunt reactive static compensator, and transformers with automatic switching under load, among others.

Most of the control devices currently in distribution networks carry out discrete control actions (TAP position change or reactive power adjustment of the capacitor banks) with large steps variations or adjustments. On the other hand, the equipment based on power electronics can be considered with continuous adjustments, which hinders the systemic VVC optimization.

4.3.1 Conventional Devices

4.3.1.1 Power Transformers with Commutation Under Load

One of the main strategies for voltage control in distribution systems is the voltage regulation through the TAPs switching (Transformer Adjust Position) of the power

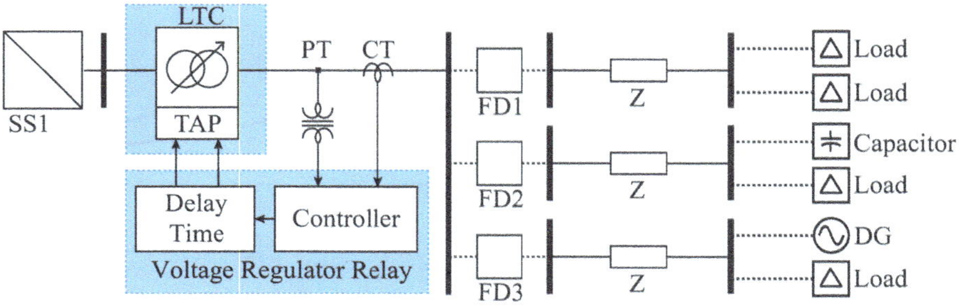

Fig. 4.4 Representation of the LTC TAP switching under the control of the voltage regulating relay

transformers with on-load tap-changer OLTC (On-Load TAP-Changers) installed in the distribution substations as shown in Fig. 4.4.

The switching of the LTC TAPs can be triggered depending on the load variations in the feeders, which modify the voltage profile in the system. The adjustment in the LTC voltage magnitude can vary in a range of $\pm 10\%$ of the voltage reference value and is realized with an Automatic Voltage Regulator Relay (AVR) or ANSI 90 function. This relay monitors the output voltage value in the secondary transformer winding and compares it with its set value, maintaining voltage control in the substation bus.

When necessary, the AVR triggers the TAP-changer to maintain the output voltage value at constant levels, even with variations in the input. Generally, the regulators used in substations are three-phase and control the three phases simultaneously. In addition, to the basic voltage regulation function, the AVR also performs protection functions for the on-load tap-changer, blocking its operation under adverse conditions, such as line short-circuit (overcurrent/undervoltage) and load protection (overvoltage).

For voltage regulation, the AVR provides six settings sets that can be adjusted: rated voltage; the timing type to start TAP switching (linear or reverse); subsequent time; type of voltage drop compensation; compensation limit; and timing step.

4.3.1.2 Distribution Transformers

The distribution transformers also have switching devices to allow the change of the windings connections, traditionally standardized with three tapping positions. However, they do not have commutation under load, requiring the transformer disconnection and the maintenance teams displacement to change them, adding costs in the system operation.

The standard values of TAP positions correspond to a difference of 5% in the TAP voltage adjustment. The transformer voltage value to the respective TAP position can be determined by

$$v_{\mathrm{TAP}_n} = v_{\mathrm{actual}} \frac{V_{\mathrm{TAP}_{\mathrm{LV}}}}{V_{\mathrm{TAP}_{n\mathrm{HV}}}} \tag{4.2}$$

where v_{actual} is the voltage measured at the equipment connection point, $V_{\mathrm{TAP}_{\mathrm{LV}}}$ the voltage of the low voltage winding, and $V_{\mathrm{TAP}_{n\mathrm{HV}}}$ the voltage of the high voltage winding.

4.3.1.3 Voltage Regulator

Voltage regulators are single-phase devices, usually installed in locations where the voltage is not compensated by the voltage drop adjustment of the distribution substation. Figure 4.5 illustrates the single-line diagram of a distribution feeder with voltage regulator across the network.

A voltage regulator is essentially an autotransformer and can operate as a voltage step-up or step-down, from the polarity of its windings, allowing voltage regulation up to ±10% of the preset limits. Typically they are constructed with eight fixed contacts (TAPs) in the series winding and one neutral position, where each contact represents 1/8 of change in the voltage regulation. They also have a reactor added to the circuit, allowing switching under load and adding intermediate voltage levels to the connections, totaling 16 voltage steps. A polarity selector switch allows you to add or subtract the voltage using the 16 steps by up to 10%, resulting in a total of 32 steps (16 steps to sum and 16 steps to subtract), each representing 5/8 of change in the voltage value.

Commonly, electronic voltage regulators have control circuits with settings that need to be parameterized, where the main ones are:

Reference voltage: is the desired voltage value at the regulator output and is expected to be maintained until a reference node, or in the load center. The output voltage can be calculated as

$$v_{\mathrm{out}} = v_{\mathrm{ref}} \frac{v_{\mathrm{pri}}}{v_{\mathrm{sec}}} \tag{4.3}$$

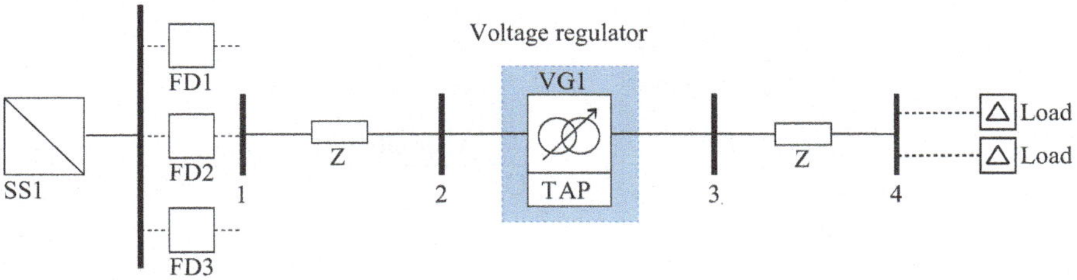

Fig. 4.5 Single-line feeder diagram with a voltage regulator

where v_{ref} is the equipment voltage within the available setting range, and v_{pri} and v_{sec} are the primary and secondary winding voltages, respectively.

Insensitivity or dead band: are the lower and upper voltage limits that guarantee the voltage regulator operation without switching.

Timing: it is the time to wait for automatic switching of the voltage regulator from the voltage violation of the set limits. There are two timing modes for the control: linear time and inverse time.

Line Drop Compensator (LDC): is a circuit that emulates the line resistance and reactance to compensate the voltage drop between the regulator set point and the load center, where the regulated voltage is desired.

4.3.1.4 Capacitor Banks

Like voltage regulators, the use of capacitor banks in the distribution system occurs both at the substation bus and along the distribution feeders. In the meantime, they are installed in shunt mode, as shown in Fig. 4.6a, providing reactive capacitive power to meet the needs of the system according to the daily load profile.

Capacitor banks can consist of several switchable modules of the same capacity, generating compensation levels, or by only one module. These modules can be controlled and adjusted both fixedly and automatically, from a sensor device that commands the operations to switch the bank on and off.

Figure 4.6b illustrates the impact of a capacitor bank with several modules on the voltage of the bus 2, signaling an increase in operating levels. It can be noted that these equipment contributes only to increase the connection point voltage, unlike the voltage regulators that allow the voltage reduction by reducing the TAP. The action strategy of a capacitor bank can be pre-established and adjustable, according to selectable limits of: time; voltage; current; or power factor.

Recently, the series capacitor installation, widely used in transmission systems, has also been explored in long distance distribution systems. The series capacitor application in the distribution system has the objective of inductive reactance

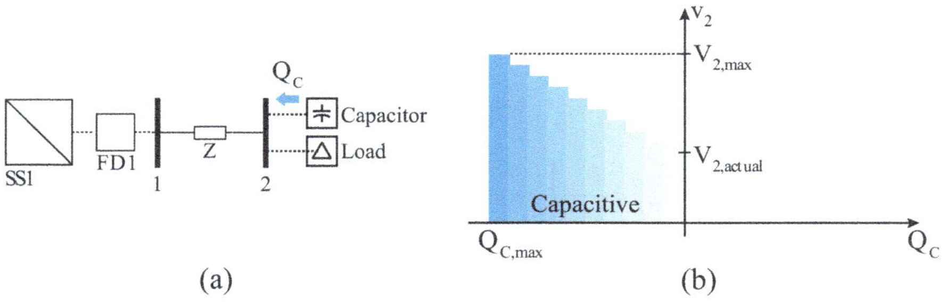

Fig. 4.6 Representation of the capacitor bank impact on the connection point voltage. (**a**) Single-line diagram. (**b**) Influence of the reactive power of the capacitor bank on the voltage level

compensation of the lines. In addition, the series capacitors can also compensate the voltage fluctuations, as well as reduce flicker effects in the system.

4.3.2 *Power Electronics Devices*

4.3.2.1 Distributed Generation and Frequency Inverters

Distributed generators connected to the distribution network, from a few kW to tens of MW, can also provide the reactive power to make the necessary voltage adjustments in the distribution system operation. To do this, one can use the frequency inverter power processing capacity, which performs the connection interface between the generation source and the electrical network.

The inverter is based on electronic components, with memory capacity, processing and communication, which allow the dc to ac voltage conversion. Actual generation sources, such as solar photovoltaic, wind power, storage systems, and also electric vehicles, operate using electronic inverters, unlike electric machines with magnetic fields that operate as generators (synchronous or asynchronous) to the active and reactive power supply, commonly used in hydroelectric generation systems.

Latest generation inverters are also known as smart inverters, and have advanced functions. Despite its converting energy capacity, they are able to respond quickly to accurate reactive power injection or absorption to compensate the energy fluctuations from intermittent sources. In addition, they do not cause harmonic distortions in the system and are able to perform more operations when compared to conventional devices, making them the ideal equipment for the modernization of the distribution networks [14].

Currently, distribution distributors in Germany and the United States are introducing smart inverters to manage the growing penetration of photovoltaic systems, since they offer new options for network operation through the inverter control and, where applicable, energy control storage [14]. Since the inverters correspond to an essential component in all photovoltaic systems, this additional control capacity, in addition to improving the existing resources, should also help in the actuations reduction of conventional voltage regulating equipment [15].

Figure 4.7a illustrates the single-line diagram of a distribution system composed of a feeder, loads, and DG. In this system, the inverter can process the active power from the generation (photovoltaic, wind, electric vehicle) by injecting it into the electrical grid to supply energy to the loads. In addition, it can enable the reactive power injection or absorption, according to the need of network voltage adjustments.

Figure 4.7b shows the influence of reactive power on the voltage levels of the DG connection point. In this case, to increase the voltage it is necessary to insert reactive (capacitive) power. The opposite occurs when it is desired to reduce the voltage, being necessary to absorb reactive power (inductive). The dc/ac converter will start operating at the instant the mains voltage is outside the proper range, following the

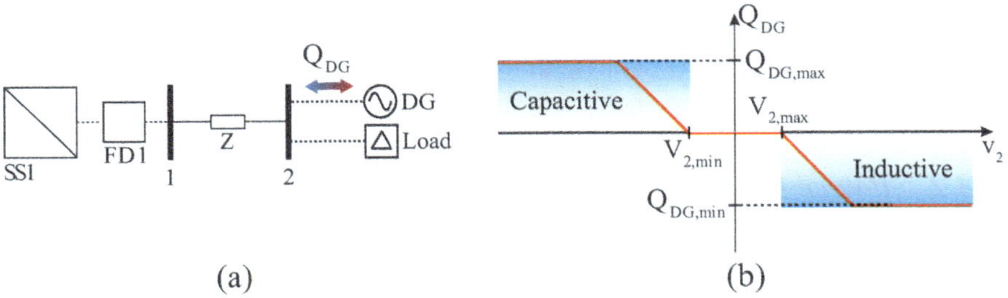

Fig. 4.7 Representation of a distributed generator: (**a**) single-line diagram and (**b**) influence of reactive power on voltage levels

characteristic shown in Fig. 4.7b. This operating characteristic is similarly applied to other recent power electronics equipment, such as static reactive compensators, more widespread in power transmission systems, and solid-state transformers.

The voltage variation in the feeder with the DGs insertion can be obtained by applying Eq. (4.4) [16].

$$\Delta V = V_1 - V_2 = \frac{R(P - P_{DG}) + X(Q \pm Q_{DG})}{V} \tag{4.4}$$

Where V_1 and V_2 represent the voltages in the initial and final distribution bus, respectively; Z corresponds to the connecting cable impedance; P and Q represent the active and reactive powers absorbed by the loads at the end of the feeder; P_{DG} is the active power generated by DG and $\pm Q_{DG}$ represents the reactive energy injected or absorbed by the DG smart inverter.

Smart inverters can operate in a variety of modes (voltage or current sources) from four power quadrants, as shown in Fig. 4.8. However, since the operational strategy used in Brazil requires that DG control only the power injected or absorbed and does not regulate the voltage of the utility system, operation in four quadrants is limited to the operation of the inverter, essentially to the active power axis with a power factor close to the unit, as highlighted in Fig. 4.8.

Although the technical and operational capabilities of smart inverters are well understood, the practical methods for configuring and implanting the devices are still limited. Continuous review of technical standards, for example, of IEEE 1547 "Interconnecting Distributed Resources with Electric Power Systems," grid codes, interconnection procedures, and communication protocols, has been directly affecting the deployment of the inverters [14]. In addition, concerns about the customer privacy and the regulatory services approval of smart inverters have also influenced the deployment strategies.

Advanced capabilities, including reactive power compensation, voltage support, frequency support for disturbances, and real-time data connectivity, can offer utilities a lower cost tool to mitigate many management challenges of smart grids. In some cases, the inverters can help to postpone or to avoid the investments in distribution, transmission, and power generation [5, 14].

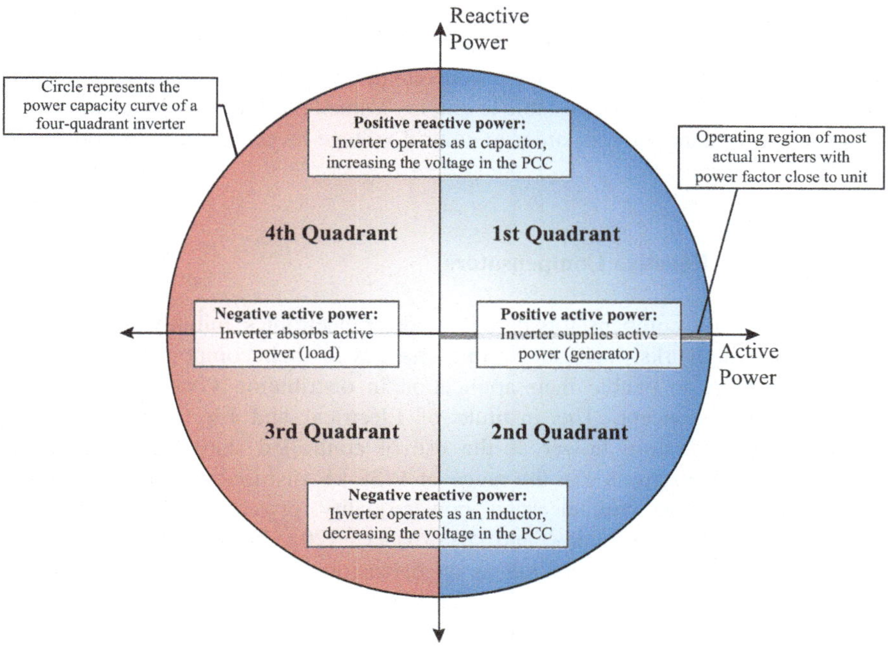

Fig. 4.8 PQ plane representation of a four quadrant inverter operation. Source: Adapted from [12]

However, there is a tendency for inverters further away from feeder transformers to be more demanding and have their life reduced. This is caused by the voltage drop and because the inverter individually controls each unit of DG.

4.3.2.2 Electrical Vehicles

Currently available Electric Vehicles (EVs) are moved by motors that are powered by electrical energy storage devices, such as batteries and supercapacitors. In general, VEs can be classified into three distinct groups: battery EV (BEV), hybrid EV (HEV), and plug-in hybrid EV (PHEV). A more comprehensive classification, called plug-in electric vehicles (PEVs), is commonly used to contemplate BEVs and PHEVs, since both need to be connected to the mains to recharge the battery.

The PEVs are presented as a new technology highly promising and with great expectation to have a fundamental role in the electric and automotive sectors in the next decades. However, it should be noted that the PEV can act as a load or power source for the electric power system. The first case occurs when the PEV is recharging its batteries, consuming power from the network. The second case can happen with the vehicle communicating with the operative center of the electric network to act for its benefit, either by providing active power or by controlling its recharge rate, a concept known as Vehicle-to-Network or V2G [17].

Through the V2G concept, PEVs can be used to provide auxiliary services such as frequency and voltage regulation in the system. Recent studies also show that PEVs can be used to inject reactive power into the electrical network [18, 19]. Thus, in the near future, it may be possible to employ a small portion of PEVs as sources of reactive power generation in order to control voltage levels and reduce grid power losses.

4.3.2.3 Static Reactive Compensators

The static reactive compensators based on power electronics, initially applied to the transmission networks, have undergone changes in their components and in the control structure to enable their application in distribution systems through the Custom Power concept. The Institute of Electrical and Electronics Engineers (IEEE) defines Custom Power as the use of controlled static-based converters based on power electronics in the range of 1–38 kV distribution voltage, in order to supply consumers with electrical power quality, protecting from momentary voltage variations, harmonics, and disturbances [20]. In addition, the Custom Power concept can be understood as an extension of the FACTS (Flexible AC Transmission Systems) concept for distribution networks, where power quality aspects become more relevant than in transmission networks.

In this sense, the voltage adjustment directly at the common coupling point of the load and the network, mainly in long feeders with rural load profile, is made possible by the Static Reactive Power Compensator for Distribution (D-STATCOM). The D-STATCOM can be series or shunt connected with the system, and has the capacity to: correct the power factor; eliminate current harmonics; and compensate for voltage drops through the inductive/capacitive reactive power processing in the distribution network [21].

4.3.2.4 Solid-State Transformer

The solid-state transformer, referred to in the technical literature by SST (Solid State Transformer), has been developed and improved over the last decades, aiming at the technical and economical viability as a promising alternative to the conventional transformer of the distribution networks [22, 23].

The SST basically consists of an ac/ac converter, with medium voltage input and low voltage output, capable to improve the power quality supplied to consumers by integrating DG sources with the grid, electric vehicles, and energy storage systems, among others, promoting the management of the energy needed in smart grids.

There is flexibility in controlling the flow of power through innumerable benefits as an advantage to the use of SST, such as: (1) the compression of sags and voltage elevations; (2) short-circuit protection with short-circuit current limiting; (3) the decoupling of the input and output voltage; (4) the correction of the power factor, as a result of the phase control of the input current; (5) regulating the output voltage;

(6) the availability of dc buses for connection purposes of distributed generation sources; (7) reactive power control and, as a consequence, the local voltage control; and (8) the bidirectional power flow.

However, when compared to the traditional transformer the SST has as main disadvantages: the reliability, the cost, and the implementation complexity.

4.4 Volt/Var Control Implementation

The structures for implementing the volt/var control are classified in the literature as local, centralized, decentralized (or distributed), and hybrid (or hierarchical) structures [6, 7]. In this sense, each control structure has advantages and disadvantages, competing the power utility to analyze its infrastructure and the desired objectives to define the best approach for the volt/var control implementation.

4.4.1 Local Control

The local control structure uses only local measurements for decision-making and considers automatic control actions in the equipment, without considering interaction between them. For this, the operative decision is based on comparisons with limit values previously calculated. Limit value adjustments require a good understanding of the system and often must be readjusted to the worst operating scenario.

Due to its rapid response, low cost, easy implementation, and also low computational requirements (no need for communication and network automation), local controls are still widely commercially used in equipment. However, because they operate based on the verification of magnitudes from the point where they are installed, and do not consider other devices in the system, they may be responsible for excessive and conflicting actions. Figure 4.9 shows a distribution system with typical equipment that can work with local controls.

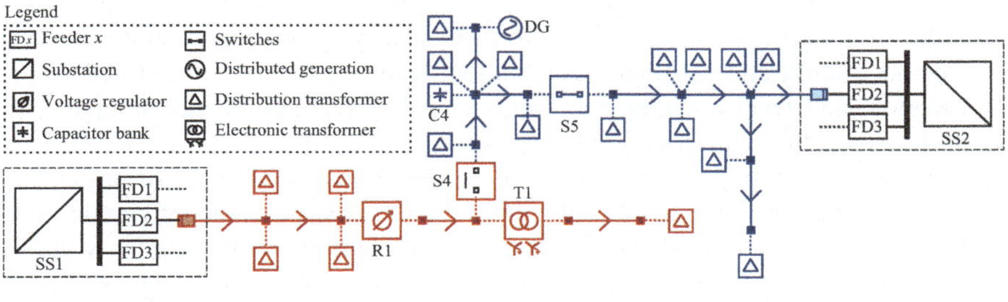

Fig. 4.9 Structure with local control devices

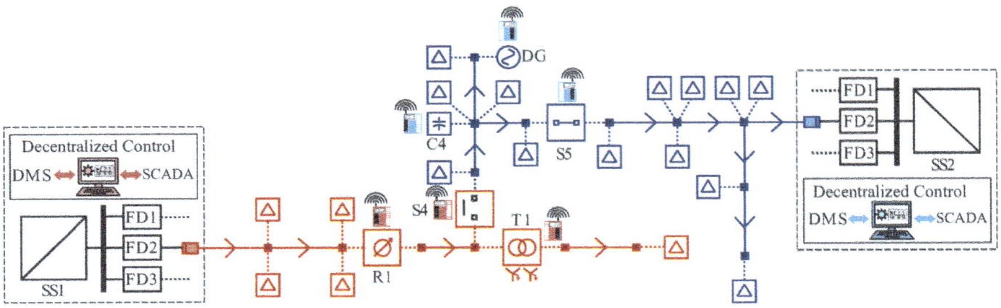

Fig. 4.10 Decentralized control structure

4.4.2 Decentralized Control

The decentralized approach refers to a set of local controllers that exchange information with a central controller in the substation, which individually controls each power supply associated with the substation.

The field equipment transmits the voltage and current data information through the Remote Terminal Unit (RTU), via SCADA communication channel, to the central control in a DMS in the substation. The central control coordinates the actions of the system devices centrally in each substation, adopting independent actions for each feeder, as illustrated in Fig. 4.10.

Decentralized control can also be referred to as centralized control at the substation.

4.4.3 Centralized Control

The centralized structure is also characterized by a central management and control system from a DMS, responsible for the decision-making functions. Modifications to controllable devices are defined for the entire utility network and not only at the level of an individual feeder or substation, as in decentralized control.

In this sense, centralized control may present more complex than the local and decentralized controls, since it covers a larger performance area. In addition, both centralized and decentralized control can be implemented in different ways, from the classification of three levels of distribution systems modernization [15]: level (a) remote control and substation automation; level (b) remote control and feeder automation, and level (c) integration and control of DG and energy storage. Figure 4.11 shows the centralized control structure.

The use of a centralized structure enables the implementation of strategies that require information at the level of the entire network, such as reconfiguration and self-healing strategies. In addition, it may also be possible to evaluate the voltages and angles states in real time, even without a large amount of measuring equipment. For this, state estimation techniques can be used.

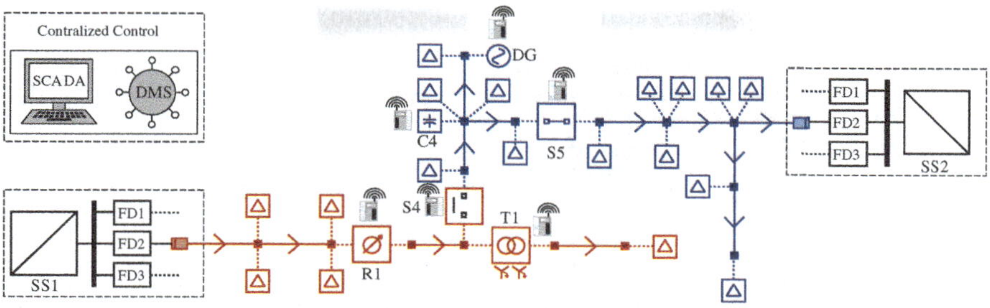

Fig. 4.11 Centralized control structure

Due to the large amount of data being transmitted to the processing and control center, a highly reliable communication system is required for high-speed and communication failures. Therefore, the loss of some measuring signal can cause an inadequate distribution system operation, being the main disadvantage of this structure.

4.4.4 Hierarchical or Hybrid Control

Hierarchical or hybrid control is a combination of centralized, decentralized, and local approaches, taking advantage of each structure [6, 7]. The hierarchical method is based on the implementation of a control scheme with the combination of the field controllers intelligence (local), the substation (decentralized or distributed), and the (centralized) utility, as shown in Fig. 4.12.

The hierarchical control structure can use different communication channels between the devices, without relying on a single communications network as in the centralized system. In addition, the hierarchical structure has the ability to integrate between the local, decentralized, and centralized controls, where lower levels of control can operate even with a communication problem. In this way, a network communication failure impacts a smaller area of the control system [6]. The hierarchical control approach is illustrated in Fig. 4.13.

The main disadvantage of the hierarchical approach is the additional hardware and software for coordinating controllable devices.

4.5 Example Case

The 10-bar test system is used to present an example case where the single-line diagram of this network is shown in Fig. 4.14a. This system has 18 loads allocated in 9 buses. In addition, two capacitor banks with 500 kVAr and with 5 switching levels

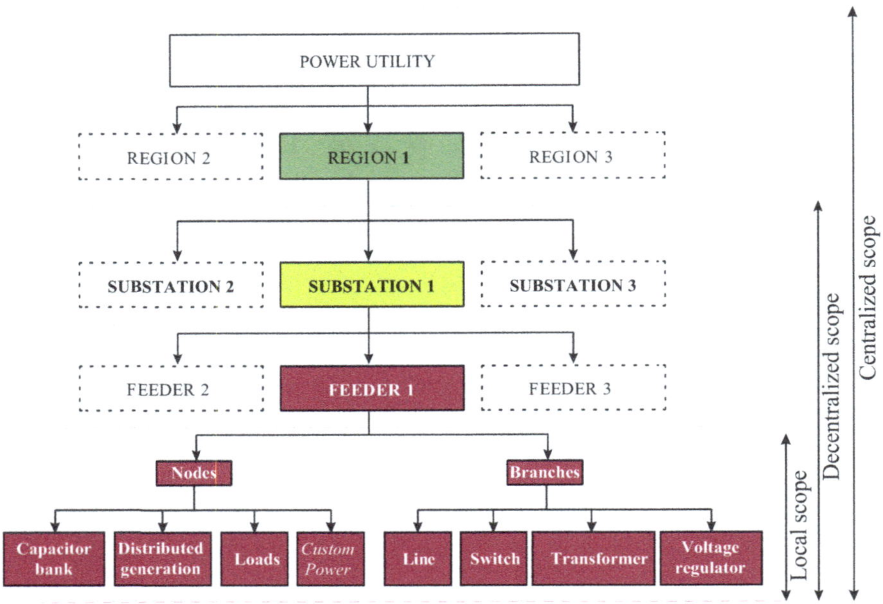

Fig. 4.12 Hierarchical VVC structure. Source: Adapted from [6]

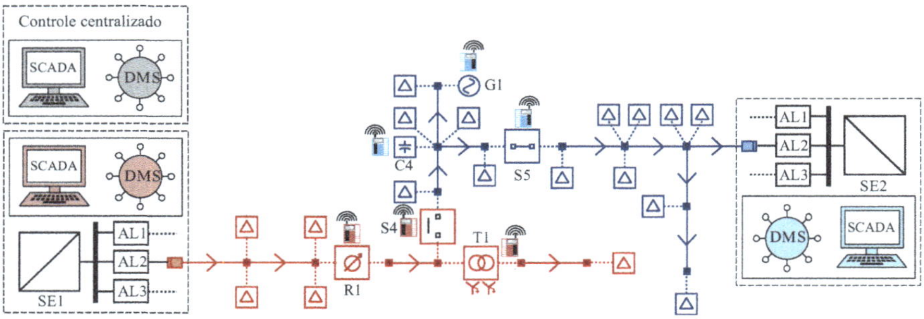

Fig. 4.13 Hierarchical control structure

(each with 100 kVAr) were considered as control equipment; and 1 grounded star voltage regulator with step of 0.00625% in ±16 setting positions.

Figure 4.14b shows the result of the voltage profile for all the bars. All control equipment is considered to be in its nominal position. It is observed that the voltage levels in some bars are outside the range considered adequate, established with voltage limits between 0.93 and 1.05 p.u.

Figure 4.15a shows the voltage result for all the bars changing the TAP positions of the voltage regulator. It is possible to note that the voltage regulator have a higher effectiveness in the downstream bars of this one, mainly due to its series connection.

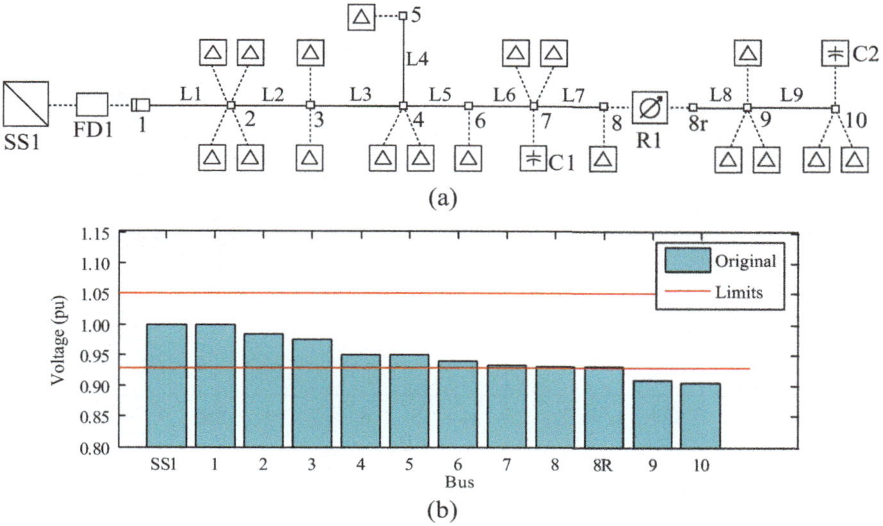

Fig. 4.14 10-bar test system. (**a**) Single-line diagram (**b**) voltage profile across the buses

Figure 4.15b shows the voltage result for all the bars changing the capacitor bank injection from 0 to 500 kVAr. The capacitor bank is a shunt-connected device, so it has a higher effectiveness in the connection point of this one (bus 10).

Figure 4.16 shows the voltage behavior across the buses before and after the VVC. The control action performed by R1 refers to the change of two TAP positions of this regulator, with 0.00625 p.u. of voltage increase by TAP. This action changed the nominal R1 condition of 1–1.0125 p.u., solving the voltage violation problem, as can be seen in Fig. 4.16.

4.6 Conclusions

The application of VVC is a way to improve the efficiency and operation of reducing power losses, maintaining a suitable voltage profile and increasing the maximum load limit supported by the feeder. However, the VVC is still little investigated in relation to power electronics based equipment, which is part of the modernization of the distribution networks and can actively help in the support to the operation of the network with actions of voltage regulation and reactive power.

For large systems, it may be more advantageous not to operate the control centralized in real time. As an alternative one can add to the centralized system small decentralized control areas, forming a hierarchical control structure. This structure provides increased reliability of conventional centralized structure.

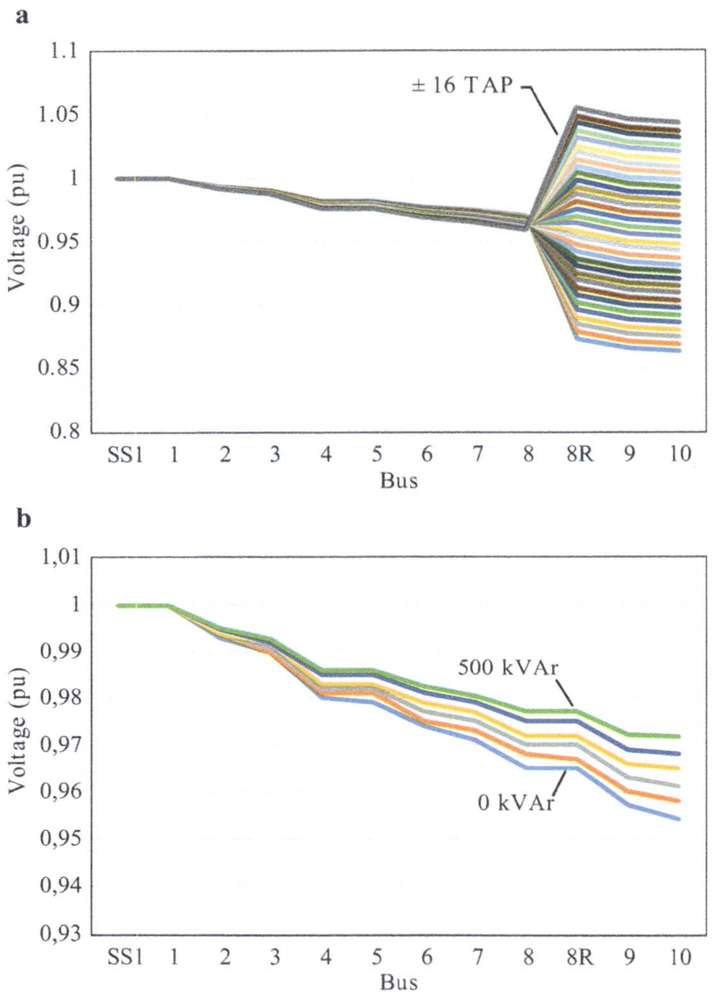

Fig. 4.15 Voltage variation caused by (**a**) voltage regulator and (**b**) capacitor bank C2

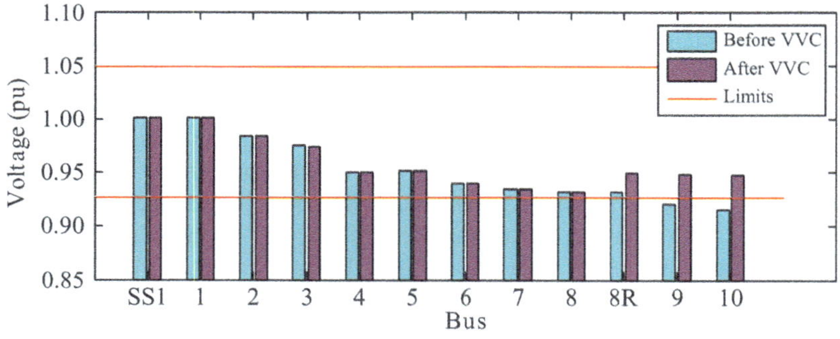

Fig. 4.16 Voltage behavior across the buses before and after the VVC

The example case demonstrated that it is possible to control the bus voltages with the VVC application, where each equipment has an effective performance in the electrical network.

References

1. Ibrahim M, Salama MMA (2015) Smart distribution system volt/VAR control using distributed intelligence and wireless communication. IET Gener Transm Distrib 9:1–12
2. Zare M, Niknam T (2013) A new multi-objective for environmental and economic management of Volt/Var Control considering renewable energy resources. Energy 55:236–252
3. Homaee O, Zakariazadeh A, Jadid S (2014) Real-time voltage control algorithm with switched capacitors in smart distribution system in presence of renewable generations. Int J Electr Power Energy Syst 54:187–197
4. Vaziri MY, Vadhva S, Tavatli F, Vaziri S (2012) Volt/VAr regulation and issues with high penetration of renewables on distribution systems. In: 2012 I.E. 13th International Conference on Information Reuse & Integration (IRI), pp. 519–524
5. Kempener R, Komor P, Hoke A (2013) Smart grids and renewables a guide for effective deployment. Int Renew Energy Agency, IRENA, Abu Dhabi, United Arab Emirates, November, 2013
6. Borlase S (2013) Smart grids: infrastructure, technology, and solutions. CRC Press, London
7. Berger LT, Krzysztof I (2015) Redes Elétricas Inteligentes—Aplicações, comunicação e segurança. LTC, Rio de Janeiro
8. Thomas MS, Arora S, Chandna VK (2011) Distribution automation leading to a smarter grid. Innov Smart Grig Technol—India (ISGT India), pp. 211–216
9. Miller L, Cibulka L, Brown M, von Meier A, Arghandeh R (2013) Distribution system voltage management and optimization for integration of renewables and electric vehicles. California Institute for Energy and Environment—California Energy Commission, Berkeley, CA
10. Markushevich N (2011) The benefits and challenges of the integrated volt/var optimization in the smart grid environment. In: 2011 I.E. Power and Energy Society General Meeting, pp. 1–8
11. Ackermann T, Andersson G, Söder L (2001) Distributed generation: a definition. Electr Power Syst Res 57(3):195–204
12. Mcgranaghan M, Ortmeyer T, Crudele D, Key T, Smith J, Barker P (2008) Renewable systems interconnection study: advanced grid planning and operations. Sandia Natl Lab Tech Rep, Oak Ridge, TN, p. 123
13. Paliwal P, Patidar NP, Nema RK (2014) Planning of grid integrated distributed generators: a review of technology, objectives and techniques. Renew Sust Energ Rev 40:557–570
14. Edge R, York B, Enbar N (2015) Rolling out smart inverters: assessing utility strategies and approaches. Sol Electr Power Assoc Electr Power Res Inst, November 4, 2015
15. Madrigal M, Uluski R (2015) Practical guidance for defining a smart grid modernization strategy: the case of distribution. The World Bank, Washington, DC
16. Ghiani E, Pilo F (2015) Smart inverter operation in distribution networks with high penetration of photovoltaic systems. J Mod Power Syst Clean Energy 3(4):504–511
17. Guille C, Gross G (2009) A conceptual framework for the vehicle-to-grid (V2G) implementation. Energy Policy 37(11):4379–4390
18. Kisacikoglu MC, Ozpineci B, Tolbert LM (2011) Reactive power operation analysis of a single-phase EV/PHEV bidirectional battery charger. In: 8th Int Conf Power Electron—ECCE Asia Green World with Power Electron ICPE 2011-ECCE Asia, pp. 585–592
19. Fasugba MA, Krein PT (2011) Gaining vehicle-to-grid benefits with unidirectional electric and plug-in hybrid vehicle chargers. In: 2011 I.E. Veh Power Propuls Conf VPPC

20. IEEE Std 1547:2003 (2003) IEEE standard for interconnecting distributed resources with electric power systems, pp. 1–16
21. Mahela OP, Shaik AG (2015) A review of distribution static compensator. Renew Sust Energ Rev 50:531–546
22. She X, Huang AQ, Lukic S, Baran ME (2012) On integration of solid-state transformer with zonal DC microgrid. IEEE Trans Smart Grid 3(2):975–985
23. Qin H, Kimball JW (Sep. 2013) Solid-state transformer architecture using AC-AC dual-active-bridge converter. IEEE Trans Ind Electron 60(9):3720–3730

Chapter 5
Protection Systems

M. A. F. Boaski, M. Sperandio, Daniel Pinheiro Bernardon, and W. S. Hokama

5.1 Introduction

The power system has undergone changes in order to provide energy to consumers satisfactorily in both technical and economic matters. Basically, transformations have arisen in all sectors by new demands, strategies, and technologies. The concepts of Distributed Generation (DG) and Smart Grids (SG) are not new, but they are now spreading. DG refers to the production of energy that is closer to or inserted in the area of consumption, which changes the model of large sources often distant from the load centers. Hence, the energy distribution networks are changing their radial operation of energy flow. On the other hand, the SG concept refers to the operation of the distribution networks with the use of technologies and methodologies to seek a more versatility structure of operation [1], which can be used for various purposes such as reconfiguration, voltage level control, and self-healing.

All of these concepts and technologies within the electricity distribution sector have the objective of increasing the network's flexibility and management capacity, thus improving the service performance to the final consumer. However, this reflects in general modifications in the subjects of study, analysis, operation, and planning [2]. These developments end up introducing new complexities in the network that

M. A. F. Boaski (✉) · M. Sperandio · D. P. Bernardon
Technology Center, Federal University of Santa Maria, Santa Maria, Rio Grande do Sul, Brazil
e-mail: mauricio.sperandio@ufsm.br; dpbernardon@ufsm.br

W. S. Hokama
CPFL Power Utility, Campinas, São Paulo, Brazil
e-mail: whokama@cpfl.com.br

D. P. Bernardon, V. J. Garcia (eds.), *Smart Operation for Power Distribution Systems*, https://doi.org/10.1007/978-3-319-93922-3_5

can have negative impacts if their operation is not adequate. Thus, there are several fields that need to be reevaluated to meet this progress, and the system's protection is one that needs a special attention.

In the electrical system it is essential to protect both the equipment and the people in cases of failures or faults in the network. However, at the same time there is a concern regarding undue operations that cause the unnecessary de-energization of consumers and the deterioration of the utility's reliability indicators. In most countries the electricity distribution networks are aerial and with bare cables, which causes greater exposure to external factors that can lead to faults, like atmospheric discharges, contact with branches, and collisions, among others [3].

Therefore, one of the main problems in the energy distribution service is overcurrents. To make an adequate protection of the networks it is essential that the equipment has the capacity to interrupt the short circuit, but one of the main points is how to assure the devices will operate when it is really necessary [4]. Hence, selectivity and coordination studies are applied to adjust the sensitivity of the current level and the correct sequence of operation of these devices.

Network protection seeks to maximize power supply and minimize areas affected by failures. As a consequence, one can verify its effect on the reliability of the system. Reliability indicators are a major concern for utilities, as they have strategic impacts on investments, as well as the financial penalties due to noncompliance with the objectives stipulated in the regulation [5, 6].

Hence, it is important to apply optimization techniques in this environment, to guarantee that the adjustments meet the needs of the system, or even to allocate the different types of devices. For that reason, there is always room for development of the techniques that can be used. Therefore, the intention of this work is to gather methods that can improve the functionality of the protection system.

5.2 Problem Formulation

The major recurring problem that power distributors face is overcurrents resulting from faults in the networks. Then, in order to avoid damage to people or the network, protection equipment is employed to detect this problem and break the circuit. However, due to the extension and branching of the distribution networks, there can be many devices in series, which need a coordinated relationship between them.

The processes that define the sequence of operation of the devices are known as selectivity and coordination. These can have a high complexity due to different types of equipment and the amount of other equipment that they relate.

To perform these procedures one of the key points is the network modeling for the protection studies. It is important to consider the proper characteristics of the distribution network, such as its topology, cables, loads, and power flow direction [7].

Therefore, the main devices involved in distribution protection will be presented, as well as the rules and standards for selectivity and coordination, in order to obtain each equipment parametrization.

5.3 Protection Devices

In the event of a fault on the network, which may result in overcurrents, overvoltages, and even a section isolation, damage may be caused to its components. This leads to a gradual or total reduction of their useful life, and exposes people to danger. The causes of faults in distribution networks can be diverse, such as atmospheric discharges, conductor disruption, contact with vegetation, collisions, and systemic instability [3].

There are several types of devices with different functionalities and applications; the complete protection of the system depends on the composition of devices. Protection has two main functions:

- Protect people, animals, and the public and private property. This function refers to the protection of the surroundings enveloped by the electrical network, avoiding damages in this environment and protecting life.
- Protect equipment installed throughout the electrical system. Besides the medium, it is important to protect the network itself and the equipment connected to it.

According to [8], it is possible to divide the devices used in the protection of electrical systems into two main types:

- Fuses: these operate by the melting of a metallic element developed with specific characteristics of time versus current.
- Relays: these represent a range of devices that can have different functions, which can monitor various electrical parameters (current, voltage, frequency, etc.) and thus provide protection not only based on current such as fuses.

As for the devices used, in recent years there has been no significant change in its operating principles, the main evolutions have been regarding the constructive characteristics, which allowed the parameterization of more refined values and often the same device encompass different functions (relays). In this sense the protection devices used in the distribution networks are fuse switches, repeater fuse switches, reclosers, and relays.

5.3.1 Fuses

The most usual protective device in overhead power distribution networks is the fuse. Mainly because of its low cost when compared to other protection devices, but it can efficiently solve one of the major problems of the distribution networks that refers to the short-circuit overcurrent. However, it is a sacrificial device; once a fuse has operated it must be manually replaced.

Being a fuse type protection equipment, it bases its operation on a metal link with specific characteristics of time versus current, that is, when the maximum tolerable

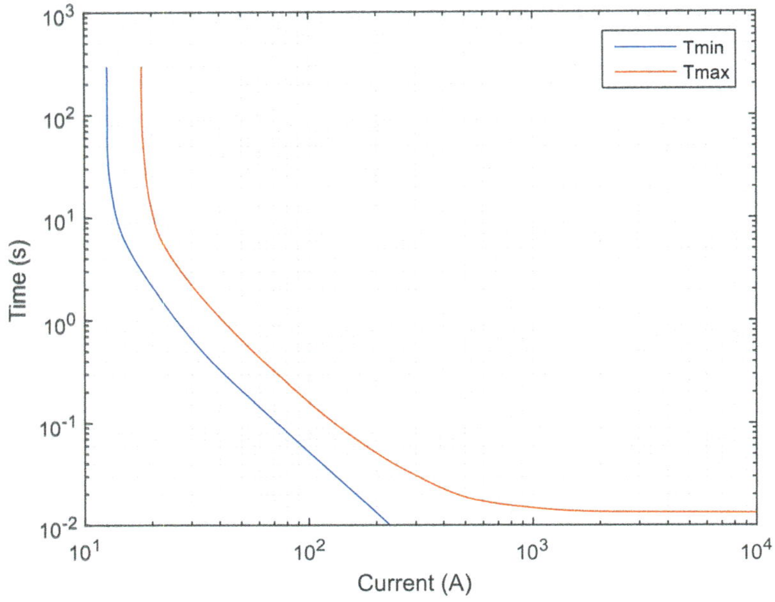

Fig. 5.1 Typical curve of fuse link—type 6 K

current is reached the heat produced melts the active element and releases the switch opening [9]. To exemplify these current versus time curves, in Fig. 5.1, the curve of a 6 K fuse link is shown. In this figure it can still be noticed that the manufacturers guarantee the performance of the links in a given area, being limited between the minimum and maximum time curves or also respectively called curves minimum performance and full recovery.

The design of a fuse is divided into three elements. Base: Consists of an insulating material (usually ceramic) and serves as an interconnection between the movable elements and the support structure of the switch. Fuse tube: consists of an insulating material, serves as a support for the fusible link and is the movable part that promotes the opening between the terminals when the link fuses. Fuse Link: consists of metal, alloys with specific characteristics of melting temperature, it is the active element of protection of the equipment.

5.3.2 Reclosers

Compared to fuses, reclosers have a relatively high cost, but these are more technological devices and offer more extensive protection, measurement, monitoring, and control [3]. They have the possibility of control and telecommunication, and through these features you can have a real-time management of the network. This allows a more complete protection and provides maneuvers for various purposes, thus contributing to the evolution of Smart Grids and Distributed Generation [10].

Today reclosers are microprocessed equipment that have a wide range of time versus current curves that can be defined through its settings, known by function 51—Time-delay Overcurrent [11]. It may also feature the instantaneous operation, known as function 50—Instantaneous Overcurrent. As an additional feature of overcurrent protection, some commercial devices rely on potential transformers or a frequency meter, which enables this equipment to perform other protection functions, such as variations in voltages or frequencies.

In addition to the aforementioned functionalities, the reclosers have a timer that allows the parameterization of automatic reclosing, without influence of the network operator. In most devices the time between new attempts is variable, being able to have fast or slow relays depending on the need of the system. The main benefit of this application is the minimization of power interruptions due to transient defects [12], which refer to defects where there is a short duration of time and thereafter the system would return to its nominal operation. In the case of transient faults and the use of reclosers, the network restoration time is negligible, and it is not necessary for a maintenance team to move for the location for repair, as it would with a fuse.

Regarding the constructive characteristics, the reclosers have basically four main parts: the input and output terminals, which are the connections that the equipment has to the mains; the current transformer (CT), which is responsible for the measurement of current parameters, and may be external or coupled to the terminals; and the extinguishing chamber, where the mobile contacts that open and close the circuit are located. Furthermore, the chamber has an environment that favors the interruption of the electric arc, which may be through vacuum, oil, or gas (the most common being hexafluoride of sulfur -SF6-). Another classification of the reclosers is with respect to its drive mechanism, which can be electric or hydraulic.

5.3.3 Relays

Nowadays, protection relays are devices that also belong to the microprocessor family. Its function is to detect faults in the network, and then command the circuit breakers to isolate the problem. The use of protection relays occurs at all voltage levels of the electrical system, from low voltage motors, generators and medium voltage networks and also on high voltage transmission lines.

In addition to the overcurrent function, which can be subdivided into a timer (ANSI table function 51) or instantaneous (ANSI table function 50), it can contain sub- or overvoltage, sub- or overfrequency functions, and differential protection between busbars, impedance, and gas pressure (Buchholz), among others.

As mentioned above, overcurrent relays are responsible for the fault detection, but these are not directly connected to the lines. Most relays have a maximum rated current of 5 A, and there are also some models with current levels of 1 A. In this way,

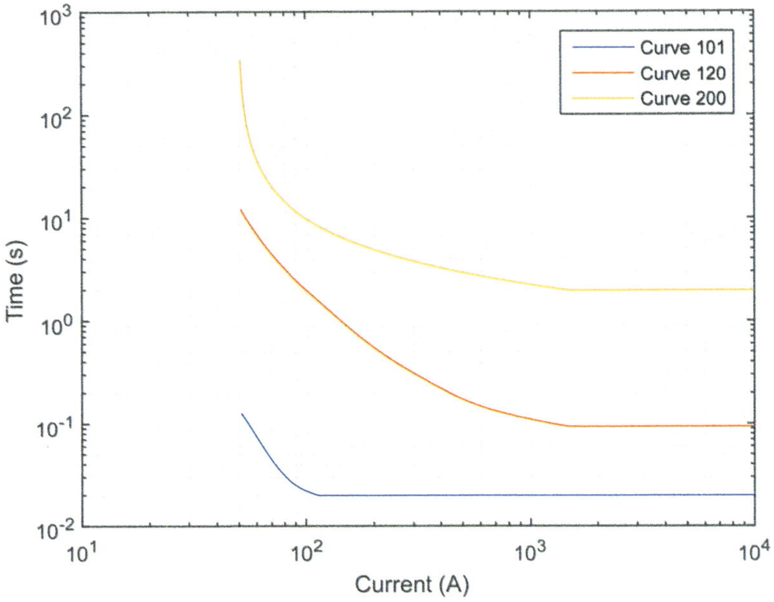

Fig. 5.2 Different types of curves

there is a need for CTs to transform the current to the appropriate levels of the protection relay.

The timed function of the protection relays is based on specific equations, and the exact types of curve of each equipment depend on the manufacturer and the model used. Some different types of curves are presented in Fig. 5.2, as an example. Another important point to be mentioned is the dial setting present in the equations that, depending on the model and the manufacturer, makes it possible to modify the curve in the time axis, maintaining the characteristic curve but changing all of it vertically, leaving it faster or slower; this is exemplified in the Fig. 5.3.

5.4 Selectivity and Coordination

Due to the importance of protection in an electrical system, in a distribution subsystem that are the feeders, there are a large number of protection devices, both for the sake of having more than one protective equipment as well as in the quest to reduce the area affected by certain fault. For a correct application of these devices there is dependence on its physical characteristics, such as interruption capacity, time of operation, communication, and others. Nevertheless, another point that needs attention is the adjustments and parametrization of devices, which refers to operation of these, searching rapid elimination of short circuits and minimization of damage [13].

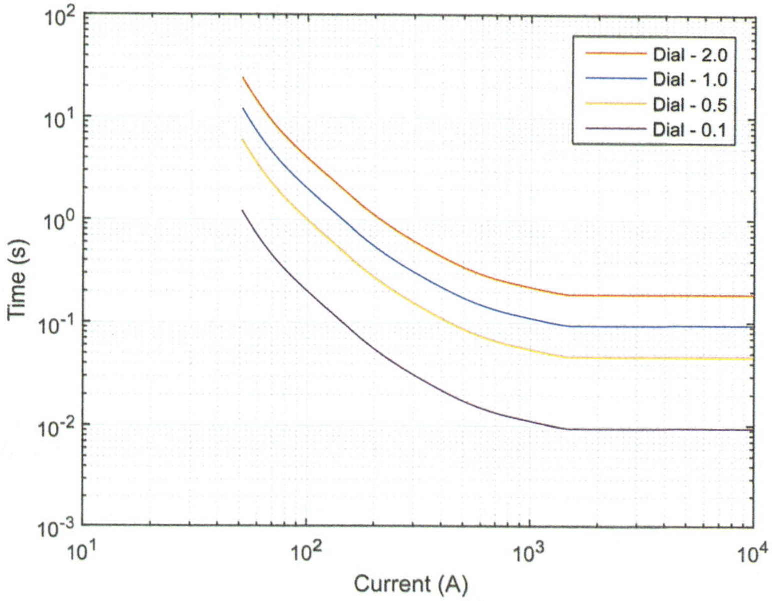

Fig. 5.3 Curve 120 with different values of dial

The adjustments and parameterizations of the equipment basically relate to the sensitivity to the current level and the operating sequence of these devices. However, this is not such a simple task, due to the large number of factors that must be considered, examples are: load current, minimum and maximum levels of short-circuit current, types of devices, downstream and upstream coordination, and operational restrictions. Then, there is a great importance involved in this parametrization, since there are a large number of protection devices in series. The improper operation of one of these equipment can cause the de-energization of consumers that should not be affected by that fault, and with that also the increase in the impact of the utility's reliability indicators [14].

There are two concepts that refer to the pairing of the protection devices, the selectivity and the coordination between devices. These will be presented separately for better understanding. Many of the criteria or standards presented may vary according to the power utility; they depend on the company policies and engineer's philosophy.

5.4.1 Selectivity

The first concept regarding the protection of electric power systems is that of selectivity. It refers to the sensitivity to the current level for its operation. In the fuse switches, the selectivity refers to the rated current of the fuse link. For microprocessed devices, such as reclosers, they refer to the pick-up current ($I_{\text{pick-up}}$), which can be adjusted according to the need of the protection specialist, but still taking into account the type, model, and manufacturer of equipment.

Many of the devices will have serial link behavior and this will apply the concepts of primary and secondary protection. The primary refers to the frontline for the fault or abnormal operation of the network; these devices should operate more quickly and disconnect only the real affected area. The secondary (rearguard or backup) protection is triggered when the primary fails to clear the abnormal situation in the system. Obviously, a proper time difference between two consecutive protections (grading time) is of essential importance for the selectivity. The triggering of the latter causes the shutdown of a larger area. The secondary protection may be only one or more devices. In this context, both equipment (primary and secondary) should be selective for short circuit. The exemplification of this process is Fig. 5.4, where both fuses (FU-1 and FU-2) should be sensitized to the short circuits in the network downstream of the FU-1.

To illustrate the question of sensitivity to the short-circuit level, the curve of a 6 K fuse link (Primary Protection) and a characteristic curve of a recloser called Kyle 120 (Secondary Protection) is shown in Fig. 5.5. For the purposes of selectivity it may be noted that there are three zones, in the left of the first vertical line (orange line) none of the devices is sensitized to currents smaller than this level. For current values that are between the vertical lines (orange and green lines), only the 6 K link fuse switch (primary device) is sensitized to these current values. Lastly, for current values greater than the second line (green line) both devices (primary and secondary) will be sensitized. Thus it becomes important to know the characteristic curve of the devices and the level of short circuit.

Another point that needs to be taken into account is in places that exists a network bifurcation, so there may be a device that is a secondary protection for two or more primary equipment. Therefore, this secondary device should consider the characteristics of two or more devices downstream. This makes the parameterization processes of protection more complex, because there can be substantial differences between the values of different regions.

For simplification of the concepts and rules in each of the equipment, each of the selectivity rules will be presented separately.

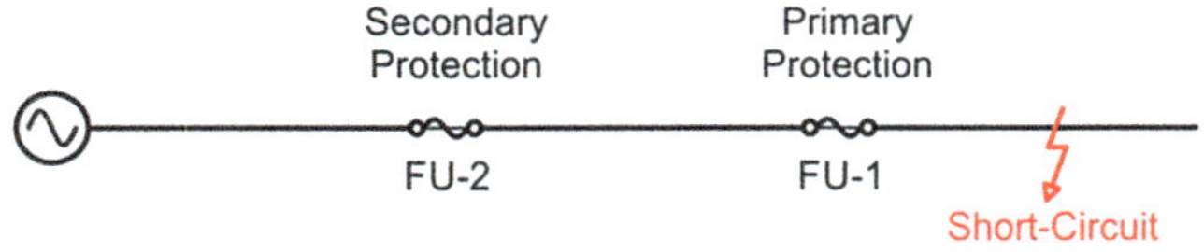

Fig. 5.4 Primary and secondary protections

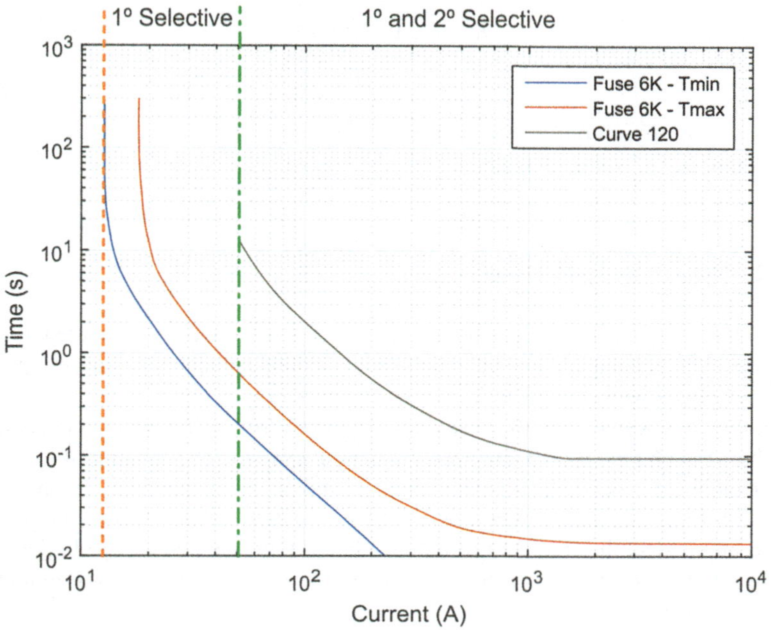

Fig. 5.5 Selectivity areas

5.4.1.1 Selectivity in Fuses

To initiate the process of the fuse melting (blows), the initial current required is approximately 1.5 times the nominal current of its respective link, for type K fuses, the most common ones along the network.

Then the selectivity in fuse has two points, the melting of the fuse link must occur for short-circuit currents, but at the same time it must allow the charging current to pass through. Thus, one can express this through Eq. (5.1).

$$I_L \cdot K_1 < I_e < I_{ccmin} \tag{5.1}$$

where:
I_L—Load current
K_1—Load growth factor
I_e—Nominal value of fuse link
I_{ccmin}—Minimun current of short circuit

The existence of this growth factor (K_1) is to have an evaluation comparing with the annual growth in a natural form of load, how long approximately this configuration will be valid without any changes. However, if there are modifications in the network, it will be necessary to reevaluate the complete study. The growth factor is defined through Eq. (5.2):

$$K_1 = \left(1 + \frac{C_{\%}}{100}\right)^{na} \quad (5.2)$$

where:
$C_{\%}$—Load growth rate of the region
na—number of years of study

5.4.1.2 Selectivity in Reclosers

The recloser's pick-up current adjustments becomes simpler compared to fuses selection. This is because they are microprocessed equipment and thus any possible value can be set between the limits of the equipment.

Another characteristic of these devices is that in addition to verifying the behavior of each of the phases, it also analyzes the neutral current. The sensing of the neutral current facilitates the fault detection that has influence of the earth. Even more in cases with fault impedance where the current variation in the phase is not so great when compared to the free short circuits. The variation is percentage higher, since the current circulating in the neutral is smaller than those of the phases, making the detection process simpler.

In this way, there will be two pick-up currents that must be parameterized in this type of equipment, phase and neutral. The selectivity in the phase adjustment is similar to that of the fuse switches. It must allow the passage of nominal load current and also consider a margin to avoid that slight variations in load cause recloser operation. Should be considered the smallest phase to phase short circuit at the end of the protection zone as the pick-up limit. Hence, the biphasic is the one of smaller amplitude. It is represented with the Eq. (5.3).

$$I_{\mathrm{L}} \cdot K_2 < I_{\text{Pick-up-phase}} < I_{cc2\phi} \quad (5.3)$$

where:
K_2—Overload factor (Generally between 50 and 100%)
$I_{\text{Pick-up-phase}}$—Phase pick-up current
$I_{cc2\phi}$—biphasic short-circuit current

The selectivity criteria for the neutral current will be different due to the behavior of the system. In order to avoid improper activation of the protection, it must be considered that in distribution systems there is always an imbalance between phases due to different loads. So, it is necessary to consider that there may be a neutral current in the nominal operation of the system, as an unbalance factor on the current load. As the upper limit for the neutral pick-up, must be considered the smaller short circuit involving earth (single-phase and biphasic to earth). There is also the possibility that these have the participation of a fault impedance, which drastically influences the final value of the short circuit. Consequently, it can be summarize in Eq. (5.4).

$$I_L \cdot K_3 < I_{\text{Pick-up-neutral}} < I_{\text{cc}+Zf\text{min}} \tag{5.4}$$

where:
K_3—Unbalance factor (generally between 10 and 30%)
$I_{\text{Pick-up-neutral}}$—Neutral pick-up current
$I_{\text{cc}+Zf\ \text{min}}$—minimum phase to earth short-circuit current

5.4.2 *Coordination*

For the good performance of the protection in the distribution systems, the initial step is to sensitize the devices against the current levels that they must operate or not. This is included in the selectivity step, but there is a concern when the sequence of operation of these equipment, dividing them into primary and secondary. In order to ensure that the sequence is met, there is a coordination of these devices. Basically, this step has the objective of evaluating the parameterizations of the protection equipment curves regarding the fault response times, verifying if there is a minimum time difference for the correct operating sequence.

This process is complex for two main reasons, the different equipment and their relationships between primary and secondary equipment. The coordinations will involve fuses and reclosing switches, and there is a diversity of protection devices types and manufacturers. Therefore, the characteristic time versus current curves of the equipment may be different. To guarantee the coordination, one must avoid the crossing of these curves and assure the minimum time difference between them. There are cases of network bifurcation, where there may be a same secondary equipment for two or more primary equipment.

The coordination process is exemplified in the Fig. 5.6. In order to ensure the correct operating sequence, the operating time of the secondary device must be longer, taking into account a minimum time difference (Δt_{min}) in relation to the primary equipment. Therefore, it is certain that the secondary device only operates after the primary protection and in case of failure of this. In this way, it is possible to visualize in the figure that for current values smaller than the highlighted value (green vertical line) the Δt_{min} is respected, ensuring the correct operation sequence. After this line, the coordination is not guaranteed, and when the curve of the secondary equipment is below the primary, it will operate first, cutting off an area greater than necessary.

Thus, the main combinations of protection devices and how the coordination between them is made will be presented.

5.4.2.1 Coordination Between Fuses

The combination of fuses with the coordination bias is somewhat simple as it has a limited number of combinations, because the operation curves are static and depend

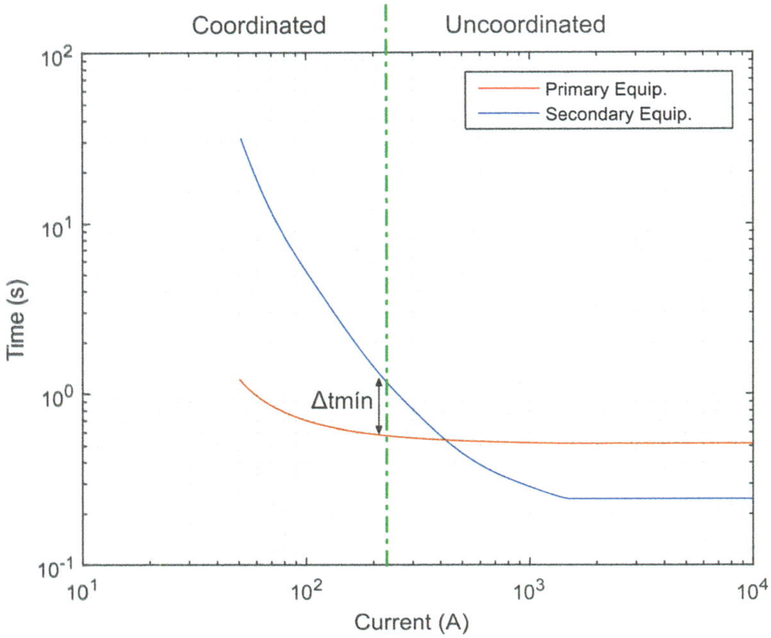

Fig. 5.6 Coordination areas

exclusively on the type and value of the link. But, sometimes, may be many fuses in series, so that the minimum time between operations is met due to this limitation [3]. In distribution network's protection, in most cases, the same type is used (Type K), because of its performance characteristics.

In order to meet the minimum time between the primary and the secondary devices, the concept used is represented in Eq. (5.5). The principle of the equation is that there will be a sufficient interval between operations if the maximum primary protection time is at least 75% of the minimum operation time of the secondary fuse.

$$T_{\max 1} \leq 0.75 \cdot T_{\min 2} \tag{5.5}$$

where:

$T_{\max 1}$—Maximum melting time of the primary fuse link

$T_{\min 2}$—Minimum melting time of secondary fuse link

5.4.2.2 Coordination Between Fuses and Reclosers

With the addition of microprocessed devices, as in the case of the recloser, the coordination process gains a greater number of combinations. This happens because that these equipment have a range of curves that can be programmed. Another point that influences the number of combinations is that the same curve can be changed through the time dial. This parameter maintains the characteristic curvature of operation but it is possible to move the curve in the time axis, being able to leave its performance faster or slower.

As the recloser curves can be changed as required, and at the same time there are many possibilities of curves and adjustments, the coordination between fuses and reclosers does not have a unique formula for their execution. What is sustained in coordination of any pair of protection devices is the need to respect the minimum time difference (Δt_{min}) between the primary and secondary operation.

An important point in the coordination that involves a microprocessed equipment is referring to the existence of the phase curve and the neutral one. Thus, both curves must respect the Δt_{min}; this is represented in Fig. 5.7. Where in relation to the minimum short circuit, there is the minimum time between the curve of the fuse and the recloser's neutral curve, while for the maximum short circuit, however, it is between the fuse and the recloser's phase curve.

As already mentioned, the possibility of combinations between operation curves increases exponentially with the use of microprocessed protection devices, in this way the scenarios for equipment coordination are broadened, not only having a single condition that meets the need for Δt_{min}. For changes between curves and equipment time dial there is no straight-line order to follow; these changes depend on the company's protection philosophy and the protection engineer's experience.

In general, a small time difference is better, as it decreases the exposure of the equipment to the fault current. However, a tight difference may incur in discoordination, because of other effects not considered, especially after a reclosing operation.

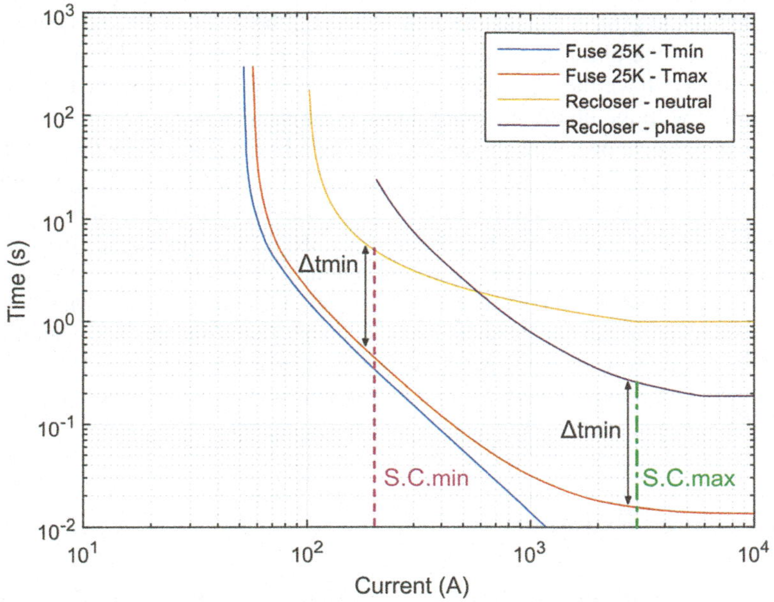

Fig. 5.7 Coordination between fuses and reclosers

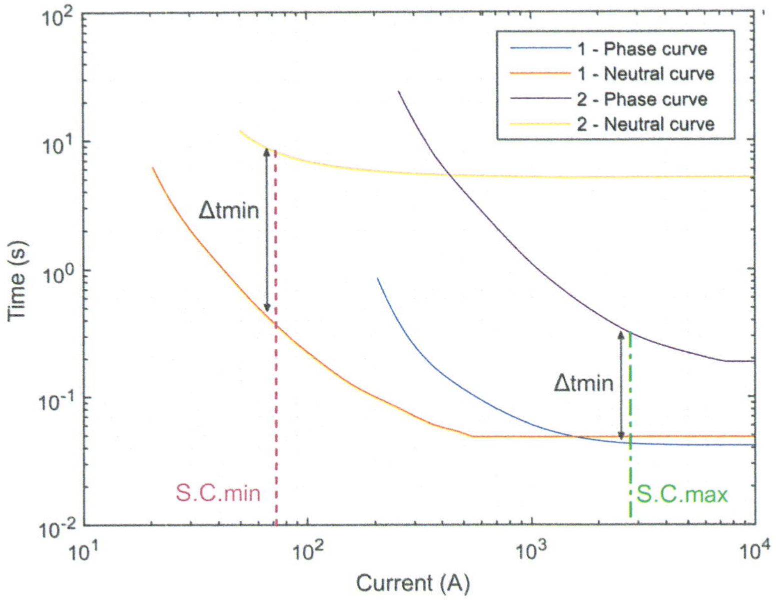

Fig. 5.8 Coordination between reclosers

5.4.2.3 Coordination Between Reclosers

In the coordination between reclosers, or circuit breakers' relays (main microprocessor devices to protect the distribution network), the analysis should consider the relationship of the phase and neutral curves of both equipment. These equipment may not always have the same characteristic curves, so it is necessary to check the combination with the phase and neutral curves of each device.

The exemplification of this process is illustrated in Fig. 5.8, where the neutral and phase curves of the primary (1) and secondary (2) are plotted. The behavior is analyzed for the minimum and maximum short circuits, verifying if along the values found it is possible to obtain the minimum time difference ($\Delta t_{\min}$) between the performances of the devices. As the coordination of these equipment depends on at least four curves, all of which may be different, an important point is to check if there is an intersection in the curves between equipment. Then, one must pay attention to the short-circuit levels of the network, verifying if there is a chance of happening an operation in that region, where the curves can intersect or $\Delta t_{\min}$ is not guaranteed.

5.5 Didactic Examples

For a better understanding of the concepts and patterns of protection of distribution systems, the relations between different devices will be presented.

5.5.1 Example 01

As the most present protection element in the networks is the fuses, the first example demonstrates this relationship. For this, the first study of selectivity and coordination consider two fuses, the circuit layout is shown in Fig. 5.9, the currents (load and short-circuit) are in Table 5.1, and the main data of devices are in Table 5.2.

The purpose of this example is to verify if the patterns of selectivity and coordination are met. In this case, check if with these current conditions, the relationship between the fuses 6 and 10 K is correct or some device should be replaced.

Thus, the first factor that will be evaluated is the selectivity, through of Eq. (5.1). As the intention is to evaluate the current condition, the load growth factor (K1) is disregarded. So, the two devices are evaluated if they are selective for the minimum short-circuit current and do not operate for the nominal load. The primary protection is a 6 K fuse, meaning its operation starts at around 12 A, which is larger than the load current on it and is also less than the minimum short-circuit current. The secondary protection link is a 10 K, so its fusion starts at around 20 A, which is higher than the load current and lower than the minimum short circuit. That is, both are selective, since the standards have been met as can be seen in Fig. 5.10.

The next step is to check the coordination, that is, if the time difference between primary and secondary is sufficient, according to Eq. (5.5). However, when checking the maximum short-circuit times in Fig. 5.10, both are operating at the same time (close to 30 ms), so does not respond to coordination. Hence, it is needed to change the secondary link to a higher value and check again all protection conditions.

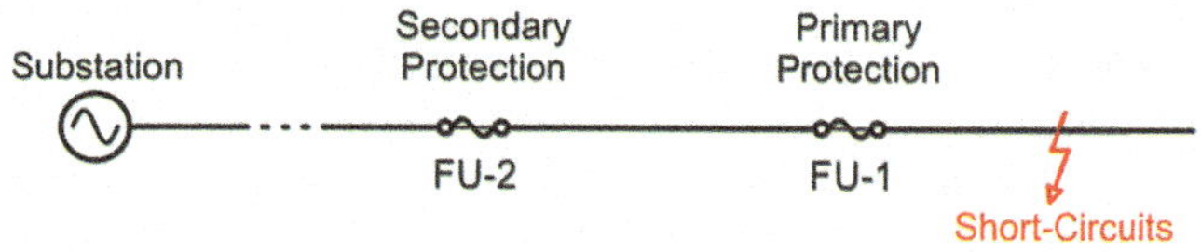

Fig. 5.9 Network from example 01

Table 5.1 Circuit data—example 01

Data	Value (A)
Maximum short circuit	250
Minimum short circuit	45
Load current on FU-1	2
Load current on FU-2	8

Table 5.2 Data—example 01

Equipment	Device type	Fuse link (K)
FU-1	Fuse	06
FU-2	Fuse	10

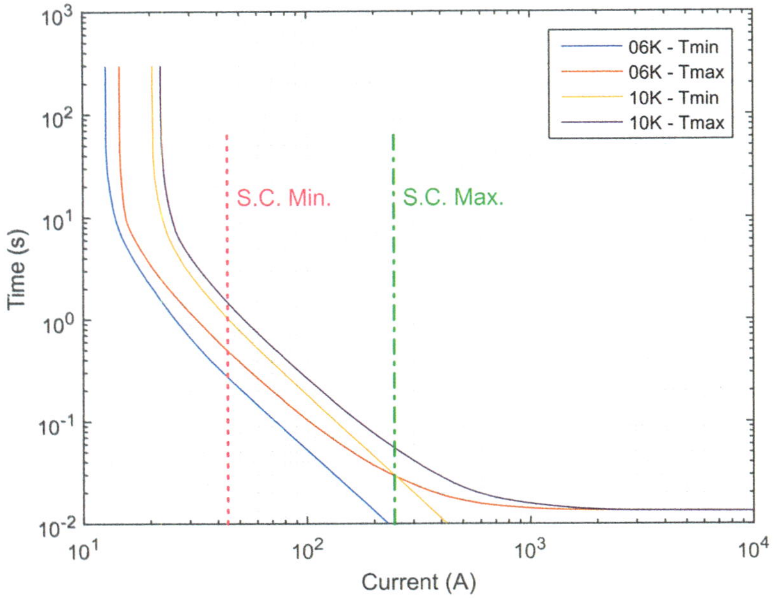

Fig. 5.10 Fuses 06 and 10 K and short-circuit currents—example 01

The next link available for the secondary equipment would be the 15 K. The selectivity of the primary device did not change in relation to that of the secondary, now its operation current is around 30 A, it is still more important that the load current is smaller than the minimum short circuit, as shown in Fig. 5.11. So, the selectivity remains.

Now with the 15 K fuse link, the operation times are: in the primary device, the maximum time ($T_{\max 1}$) of operation is 30 ms, and in the secondary, the minimum time ($T_{\min 2}$) is 80 ms. With these operating times, Eq. (5.5) is satisfied, so the system is selective and coordinated.

5.5.2 *Example 02*

As a second example, is system with 03 equipment in series, as presented in the Fig. 5.12. When there are more than two devices in series, the most used process of selectivity and coordination of protection occurs through pairs of devices. In this case, two pairs will be performed, the first with the fuses (FU-1 and FU-2) with the short circuits of area "A" and the second is the fuse (FU-2) and recloser (R-1), with the short circuit of region "B." The current data is in Table 5.3 and the data of devices are in Tables 5.4 and 5.5.

In order to do the selectivity and coordination analysis of the protection of this example, they will then be separated by pairs of equipment, the first pair is

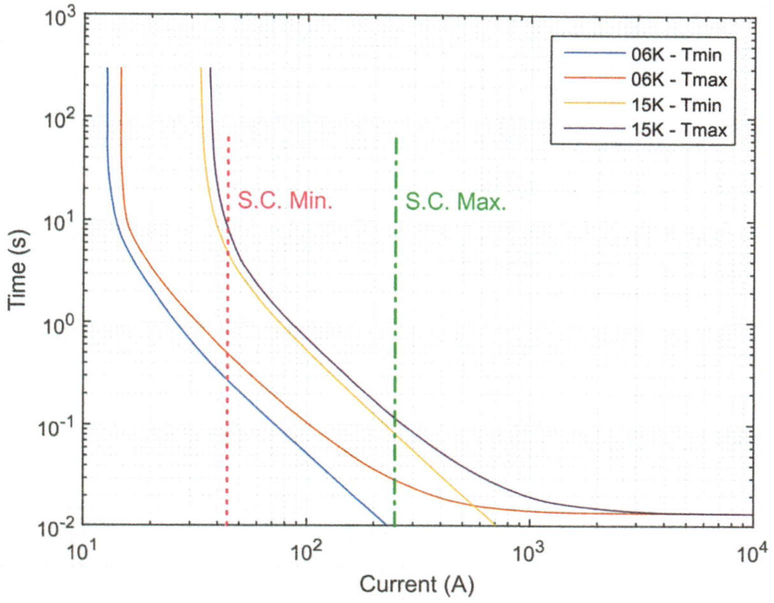

Fig. 5.11 Fuses 06 and 15 K and short-circuit currents—example 01

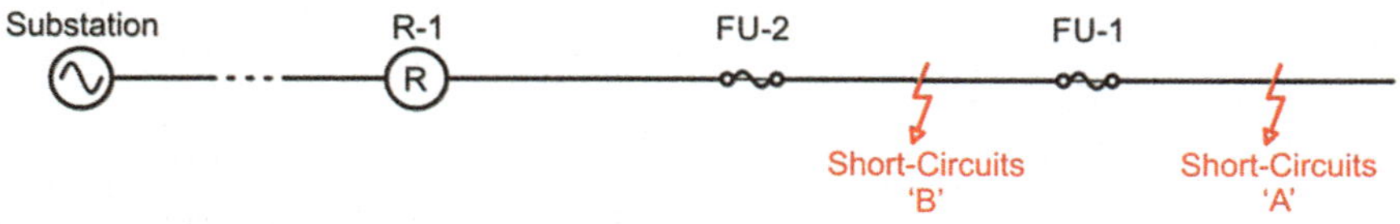

Fig. 5.12 Network of example 02

Table 5.3 Circuit data—example 02

Data	Value (A)
Maximum short circuit "A"	418
Minimum short circuit "A"	55
Maximum short circuit "B"	485
Minimum short circuit "B"	70
Load current on FU-1	2
Load current on FU-2	8
Load current on R-1	

demonstrated in the Fig. 5.13. The question of selectivity with the minimum short circuit is presented in Fig. 5.14, and both are sensitized.

The coordination that requires time analysis in the maximum short-circuit current too is shown in Fig. 5.14. The minimum operation time of the 25 K link is approximately 80 ms, and the maximum link time of 06 K is approximately

Table 5.4 Fuses data—example 02

Equipment	Device type	Fuse link (K)
FU-1	Fuse	06
FU-2	Fuse	06

Table 5.5 Recloser data—example 02

Data	Value
Device type	Recloser
Phase curve	120
Neutral curve	137
Ipk-p	150
Ipk-n	60
Phase dial	0.2
Neutral dial	0.8

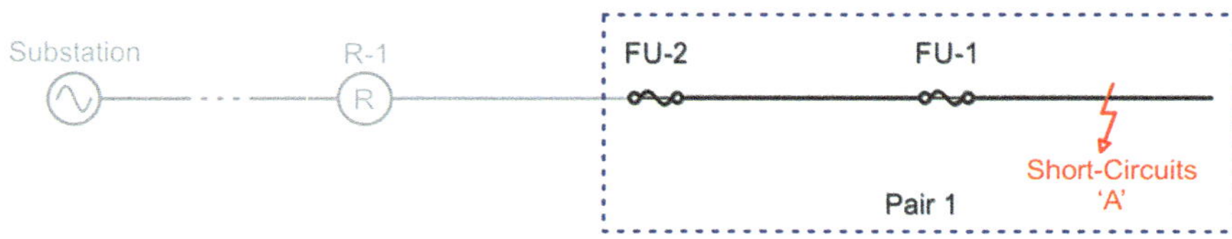

Fig. 5.13 First pair of devices—example 02

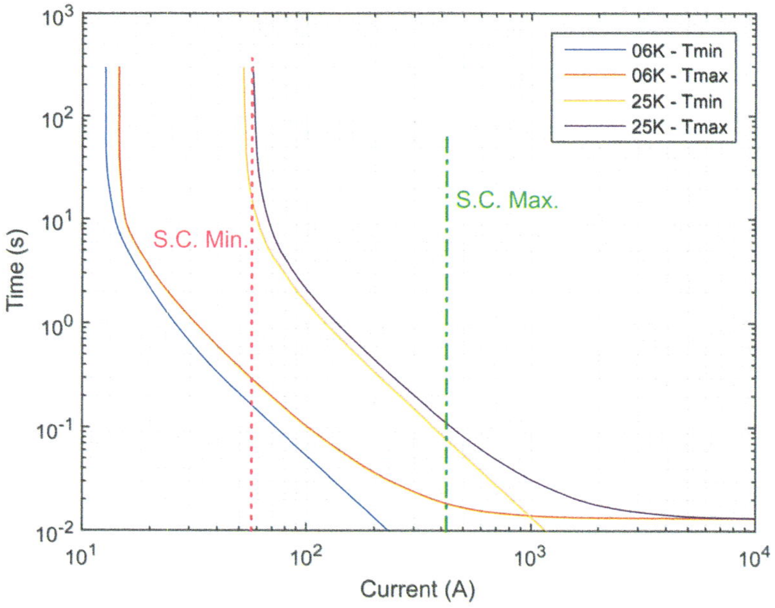

Fig. 5.14 Fuses 06 and 25 K and short-circuit currents—example 02

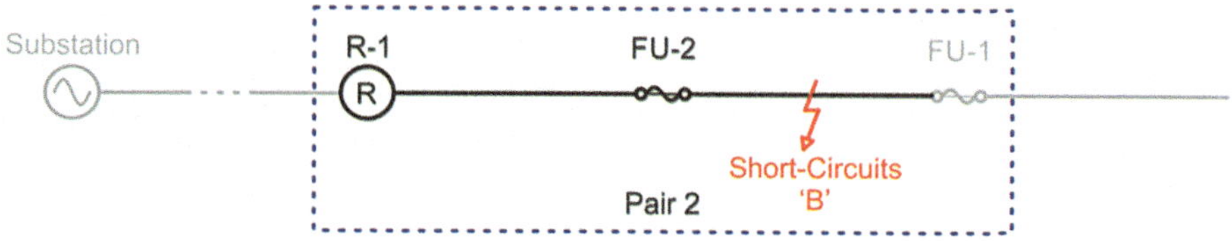

Fig. 5.15 Second pair of devices—example 02

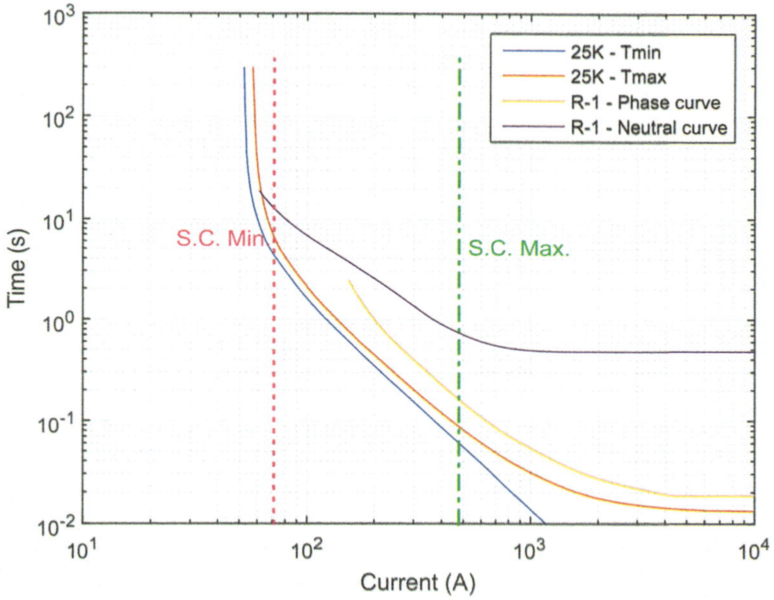

Fig. 5.16 Fuse 25 K, recloser curves and short-circuit currents—original—example 02

20 ms. In this way, it follows Eq. (5.5), which means that these equations are coordinated.

The second pair refers to FU-2 and R-1 as shown in Fig. 5.15. In this case, as there is a recloser involved, the test for selectivity could be based only on the pick-up current, since it starts the equipment operation. Nevertheless, it can be verified through Fig. 5.16 that both are sensitized to the minimum short circuit of area "B."

As for the coordination of a microprocessed device, the relation is not through Eq. (5.5), only a desired Δt_{min} is expected. In this example the minimum difference time of 300 ms will be considered as coordinated. To carry out this analysis is considered the maximum short-circuit current of Fig. 5.16, so that a time of approximately 90 ms is found for the fuse and 160 ms for the recloser, which has its current level in this current through the phase curve. However, these configurations do not meet the minimum time difference. Considering the possibilities of the recloser, one adjustment is in the phase curve dial, which is the point that will be changed to meet the coordination.

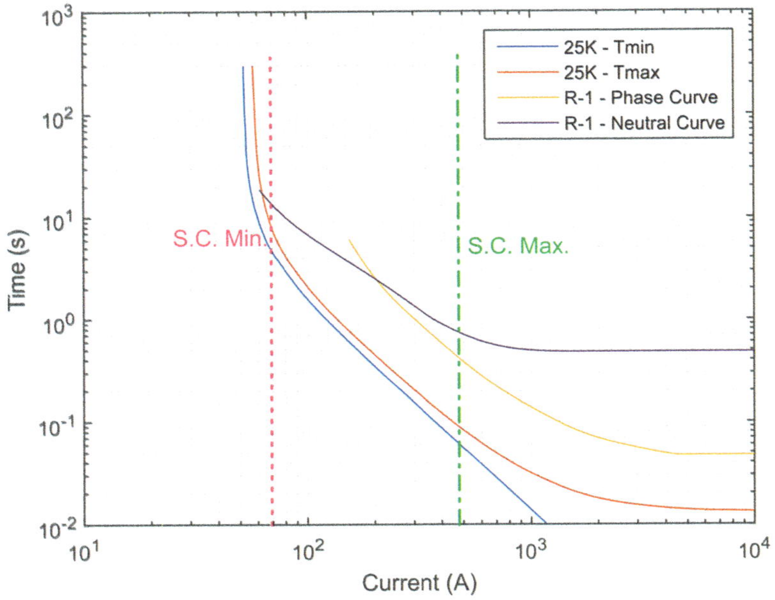

Fig. 5.17 Fuse 25 K, recloser curves and short-circuit currents—modified—example 02

With the change of dial of the curves, it does not influence the selectivity of the device; there is only a vertical variation of the curve, so it is not necessary to reevaluate the selectivity. Then it will only be reevaluated as to coordination. In this case the phase dial was changed to 0.5, and the behavior is shown in Fig. 5.17. With this modification, the recloser's operation time was 400 ms and the fuse switch was kept at 90 ms, taking into account the minimum time difference stipulated and achieving coordination between the devices.

5.6 Smart Protection

The power system today has an increased complexity with the integration of distributed energy resources (DER), which includes embedded or distributed generation (DG) and electrical vehicles (EV), challenges the handling of faults because of the changing in the power flow [15].

This has a strong impact on protection of distribution system, but also on the transmission level. In the distribution system, directional overcurrent or distance relays becomes required, and strict voltage and frequency protection have to be relaxed [16]. In the transmission system, changes in power flow direction could pose challenges for backup protections.

The newest generation of multifunctional relays has characteristics and behavior that can be programmed, and are able to communicate in real-time, using information from other parts of the power system. This technology can be smartly used to ensure the system's adaptability to correct disconnection of faulty items in order to minimize the consequences as much as possible, increasing the utilization of the power grid with reduced risk of blackouts.

Therefore, a Smart Protection must seek to improve classical power system relaying by utilization of communication technology. The aim is increased reliability, with better handling of faults and transients caused by DER's dynamic that can influence relay performance.

The smart grid technologies to enable these new protection strategies involve:

- Investigate and demonstrate new protection and fault location methods for power systems with DER
- Develop and test new distributed sensor technology for fault location and handling
- Develop robust and flexible communication architectures and protocols
- Utilize synchrophasor technology to implement and demonstrate wide area protection and control

All of these must be integrated by an Advanced Distribution Management System (ADMS), which is a software platform with numerous utility systems that provides automated routines without paperwork, manual processes, and siloed software. An example of this concept is illustrated in Fig. 5.18.

Rating and technology of the DER units are important factors, especially their power electronics interfaced units. The coordination of voltage and frequency protection with the feeder protection is of particular interest, and distance relays must be considered in distribution grids.

The communication's usage have been more focused on slow-changing system operations or monitoring applications. However, a fast data communication is now needed with distributed sources units and loads for optimum protection and fault location.

A wide-area protection and control is required to integrate protective relays, transient event recorders, and dedicated PMU devices, providing and sharing synchrophasor data, essentially GPS time-tagged phasor measurement information.

This structure can make feasible: high-speed emergency control against power swings or voltage collapse [17]; applications for unbalanced disturbances, for example, single-pole tripping and reclosing [18]; smart high-speed separation and optimal islanding when emergency control cannot hold the whole system together [19]; operation of separated islands as microgrids [20].

Nevertheless, the application of this concept may run into interoperability issues, since many manufacturers still practice proprietary communication protocols. Hence, the protection systems should be based on the IEC 61850. It is an international standard defining communication architecture for electric power systems. The abstract data models featured in it can be mapped to a number of protocols.

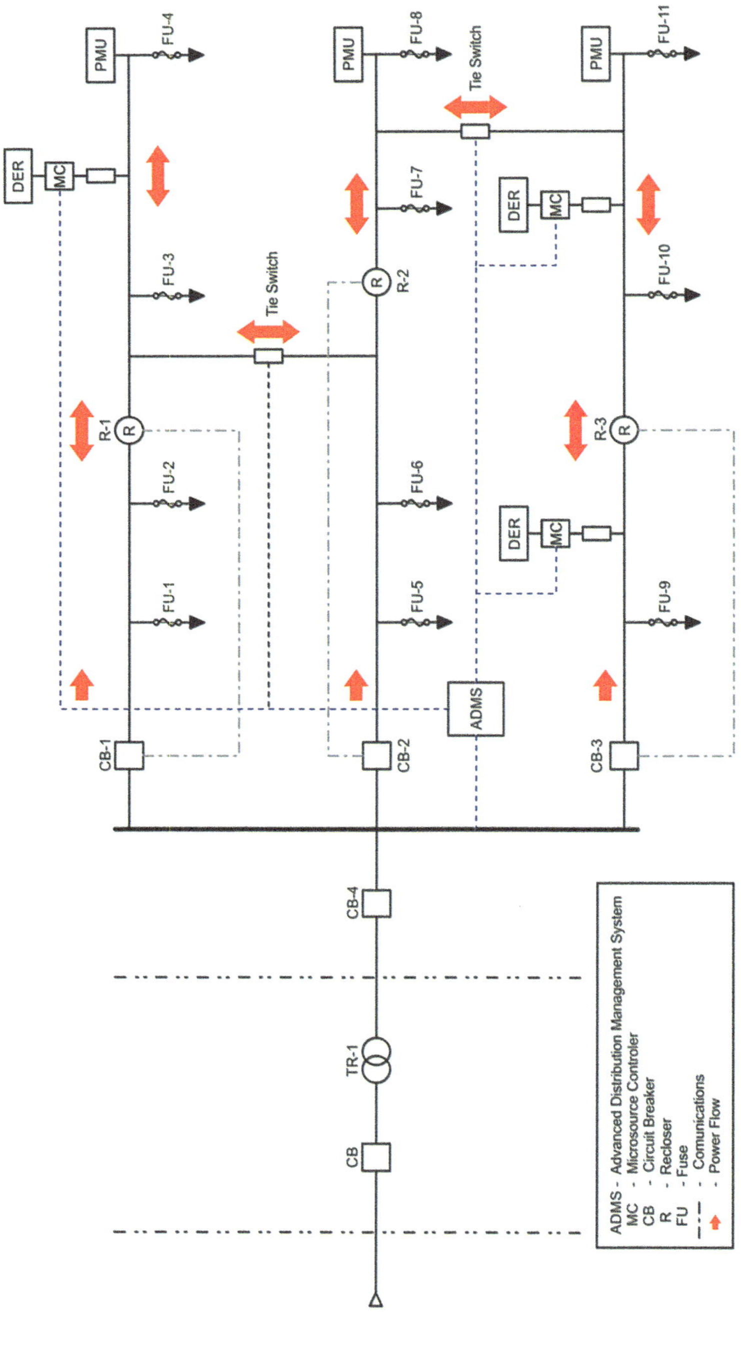

Fig. 5.18 Advanced distribution system management

The current standard relies on a homogenous communication system where one transmission reaches all possible recipient [21]. The Quality of service (QoS; delay, packet loss and delay jitter) is almost constant with no variation between different sources and receivers. Extending this beyond the substation (a single Ethernet) is challenging, since these assumption do no longer apply. The QoS will be heterogeneous and vary in both space and time. Distributed control and distributed sources imply dynamic configuration of the flows of control and data. In a wide area multicast setting, there is always a trade-off between robustness, load, and reliability of packet delivery, delay and delay jitter. The challenge is to design network services with sufficient QoS for the new real-time control algorithms that will scale beyond the substation. The cyber security aspects are important for multicast systems, which are vulnerable for misconfiguration, attacks, and manipulation.

It is essential to confirm the functionality of an overall integrated system, with an optimal data management and availability to real-time distributed control processes. It is necessary to investigate beneficial system information structures and algorithms from centralized to distributed or combined hierarchical, like robust unicast and multicast architectures.

The ethernet connectivity to protection brings cybersecurity measures that can affect the real-time performance, so algorithms for optimal balance between performance and security must be developed. Data overload will occur and the "big data" problem needs solutions.

5.7 Conclusions

In this chapter, the main characteristics of the protection systems for distribution networks were presented, as its coordination and selectivity procedures. A short time ago, all these procedures have to be made manually by engineers, based on data not constantly updated. Nowadays, this is incompatible with the variability of DERs and load profiles, the possibility of "on-the-fly" network reconfiguration, and the necessary reliability.

It is possible to visualize the functional characteristics of the protection devices, in which it can be noticed that the fuses have their actuation curve connected to the installed link; already equipment as relays and reclosers, they have a greater flexibility in its operation, with different types of curves and with the possibility of varying them. But besides the use of equipment with capacity, it is important the relationship between the existing equipment in the network, and for this there are the concepts of selectivity and coordination that aim to control the sensitivity to current levels and also the order of their performance compared to short circuits. The examples contributed to the visualization of the presented concepts besides demonstrating how adjustments can be made when the established standards of selectivity and coordination are not met. In addition, it contributed to the validation of the importance of coordination and selectivity of the protection devices of an energy distribution network.

References

1. Brown RE (2008) Impact of smart grid on distribution system design. IEEE Power Energy Soc 2008 Gen meet convers Deliv Electr energy 21st Century, PES
2. Bernardon DP, Mello APC, Pfitscher LL, Canha LN, Abaide AR, Ferreira AAB (2014) Real-time reconfiguration of distribution network with distributed generation. Electr Power Syst Res 107:59–67
3. Ramos MJS, Bernardon DP, Comassetto L, Resener M (2013) Analysis of coordination and selectivity of protection systems during reconfigurations of distribution energy systems in real time. In: 2013 I.E. PES conference on innovative smart grid technologies, ISGT LA. IEEE, São Paulo, Brasil, pp 1–6
4. Comassetto L, Bernardon DP, Canha LN, Abaide AR (2008) Software for automatic coordination of protection devices in distribution system. IEEE Trans Power Delivery 23:2241–2246
5. Siirto OK, Safdarian A, Lehtonen M, Fotuhi-Firuzabad M (2015) Optimal distribution network automation considering earth fault events. IEEE Trans Smart Grid 6:1010–1018
6. Xu X, Mitra J, Wang T, Mu L (2016) An evaluation strategy for microgrid reliability considering the effects of protection system. IEEE Trans Power Delivery 31:1989–1997
7. Ledesma JJG, De Araujo LR, Penido DRR, Pereira JLR (2015) Algorithm for the analysis of coordination of protection for distribution systems. IEEE Lat Am Trans 13:2168–2175
8. Blackburn JL, Domin TJ (2014) Protective relaying: principles and applications, 4th edn. CRC Press, Boca Raton
9. Khorshid-Ghazani B, Seyedi H, Mohammadi-ivatloo B, Zare K, Shargh S (2017) Reconfiguration of distribution networks considering coordination of the protective devices. IET J 11:82–92
10. Velásquez MA, Quijano N, Cadena AI (2016) Multi-objective planning of recloser-based protection systems on DG enhanced feeders. In: 2015 I.E. PES innovative smart grid technologies Latin America, ISGT LATAM, vol 2015. IEEE, Montevidéu, pp 1–6
11. IEEE (2008) Standard electrical power system device function numbers, acronyms, and contact designations. In: IEEE Std C37.2–2008, pp.1–48
12. Jamali S, Borhani-Bahabadi H (2016) Recloser time–current–voltage characteristic for fuse saving in distribution networks with DG. IET Gener Transm Distrib 11:272–279
13. IEEE (2001) 242–2001 I.E. recommended practice for protection and coordination of industrial and commercial power systems
14. Comassetto L, Bernardon DP, Canha LN, Abaide AR (2008) Automatic coordination of protection devices in distribution system. Electr Power Syst Res 78:1210–1216
15. Liu Z, Su C, Hoidalen HK, Chen Z (2016) A multiagent system-based protection and control scheme for distribution system with distributed-generation integration. In: IEEE transactions on power delivery, vol. 32 (1)
16. Pandakov K, Høidalen HKr, Marvik JI (2016) Implementation of distance relaying in distribution network with distributed generation. In: 13th Int Conf on Development in Power System Protection, Edinburgh, UK
17. Tziouvaras D (2007) Relay performance during major system disturbances. In: 60th Annual Conference for Protective Relay Engineers, College Station, USA
18. Calero F, Hou D (2006) Practical considerations for single-pole-trip line-protection schemes. In: 5th Annual Clemson University Power Systems Conference, Clemson, USA.
19. Isazadeh G, Kordi M, Eghtedarnia F, Torkzadeh R (2017) A new wide area intelligent multi-agent islanding detection method for implementation in designed WAMPAC structure. In: 4th International Conference on Power and Energy Systems Engineering, Berlin, Germany
20. Salazar J, Tadeo F, de Prada C, Palacin L (2010) Modelling and control of microgrids in island operation, the International Renewable Energy Congress, Sousse, Tunisia
21. Adrah CM, Bjørnstad S, Kure Ø (2017) Fusion networking for IEC 61850 inter substation communication: use case applications. J Commun 12:9

Chapter 6
Operations Planning

L. F. Braghirolli, Vinícius Jacques Garcia, and J. S. Fonini

6.1 Introduction

This chapter presents the main elements related to the operation responsible for satisfying the requests demanded by the customers of an electricity distribution system operator (DSO). Usually, the focus of the energy distribution planning is on the technical aspects of the distribution network [1, 2], with little attention on the planning required for the teams to act efficiently in the grid expansion, maintenance, and customer-related services [3]. The objective is to present the main drivers of planning for these services in the long-, medium-, and short-term horizons.

Distribution is the last stage of the energy supply chain, after generation and transmission. The difficulty for two companies to provide energy to a particular customer makes this market an example of a natural monopoly. However, such monopolies must be regulated in order to guarantee the efficiency and quality in serving the population [4]. Among the mechanisms used by the regulatory agencies is the benchmark, which sets the standards for rewarding efficient companies and allows the regulatory agencies to establish fines to penalize inefficient ones [5–7].

The insertion of new technologies for distributed generation can impact the way the energy distribution is managed and how the DSOs are organized [8]. One

L. F. Braghirolli (✉) · V. J. Garcia
Technology Center, Federal University of Santa Maria, Santa Maria, Rio Grande do Sul, Brazil
e-mail: lynceo.braghirolli@ufsm.br; viniciusjg@ufsm.br

J. S. Fonini
RGE Sul Power Utility, São Leopoldo, Rio Grande do Sul, Brazil
e-mail: julio.fonini@rgesul.com.br

D. P. Bernardon, V. J. Garcia (eds.), *Smart Operation for Power Distribution Systems*, https://doi.org/10.1007/978-3-319-93922-3_6

tendency is to increase competitiveness in the services related to distribution [9], including those carried out by the teams along the grid, being able to arrive at a complete restructuring of the current business model through servitization [10]. In this changing environment [8, 9], the improvement of service operation planning and management is consistent with future needs, allowing companies to begin to develop the skills to operate in a more competitive context. The service management factors discussed in this chapter are fundamentally related to the need to manage geographically distributed requests. The primary focus of this chapter is the integration of the requirements imposed by the business context (types of services, customer criticality, regulatory agency penalties, etc.) in the process of planning the field activities of the teams.

In the next Sect. 6.2, we present the types of services that are carried out in the field by the teams. This characterization allows a better understanding of the conditioning factors and the demands that the operation must meet. Section 6.3 describes the resources that DSOs typically use to service the requests. The characterization of the central decisions in the planning of the service operation is presented in Sect. 6.4. Finally, the conclusions are presented in Sect. 6.5.

6.2 Types of Services

The services performed by the teams as part of electricity distribution operation have different natures. Three groups of services require the displacement of a team to the place in the network where the service should be performed. The first group is related to the expansion of the system. New network branches are needed to develop previously unpopulated regions resulting from urban sprawl. The increase of the reliability of the system also depends on new network branches, required to increase the points of maneuver. Another example is the increase in demand, which implies the need to increase the capacity of conduction and transformation in the electric distribution network.

The second group of services is related to the maintenance of the system. Predictive maintenance activities involve monitoring the behavior of the network, either through system collected data or inspections. An example of this type of service is the replacement of wooden poles in the early stages of decay. Natural decomposition increases the likelihood of network failure and can result in a hazardous situation to the population. It should be contained whenever situations close to critical levels are identified. Preventive maintenance corresponds to the services demanded as established in a maintenance program, anticipating the failures and occurring regardless of the degree of wear of the component. The replacement of transformer oil in substations is an example of this type of activity. Finally, the corrective maintenance represents the response of the teams to the failure of a component. In this case, the power supply interruption occurs, making the service extremely critical. This type of situation is not desired and usually characterizes an emergency situation to restore the supply to the customers.

The third group contains the services related to the needs of customers regarding the supply of electricity. In this group, there are services with different characteristics, but all have their origin in the client. Examples of this type of service are the connection of a new customer, disconnection when the contract expires or due to lack of payment, maintenance to the customer's meter, and changes in supply (increase/load reduction), among others.

When managing such services, it is necessary to consider some aspects that will determine the degree of importance and the urgency of each one. The safety of the population and the customers served is a fundamental aspect, since a failure in the distribution system can put several people at risk. For example, two corrective maintenance requests may have different priorities depending on the risks inherent in each situation. A customer without power due to a failure of the meter terminal connection presents a lower risk than another customer without power due to a broken wire that is still energized. In this case, the second request has higher priority due to the associated risk, and the teams should perform the second request first and as soon as possible.

The type of affected client is also an essential criterion in identifying priority requests. In general, clients can be characterized by the importance it represents for society. Critical clients carry out activities whose interruption of energy supply can lead to irreversible damage to society, such as hospitals and water treatment plants, which are essential services that serve large numbers of people. Relevant customers carry out activities whose disruption affects not only who hires the energy supply but also a larger group of people, such as in industries, commerce, schools, and public services in general. Regular customers comprise all other individuals who contract the power supply, and whose interruption of supply has impact only on the contractor. In this category are mainly residential customers.

Another aspect that influences the management of these requests is the number of clients affected. The same type of fault occurred at two distinct points in the network can have a significantly different impact. For example, a fault in a transformer can reach 2000 residential customers if it happens at one network point, while it can affect only 10 customers if it happens at another point. In this case, the priority of the first request is higher, because the same work of the teams can meet 2000 clients at once. The failure impact can also be inferred as a function of the interrupted load. It is assumed that customers with higher energy consumption have a more significant social impact and, therefore, should be met more quickly. Both the number of customers affected and the interrupted load (associated with the type of customer) are factors usually considered by the regulatory agencies in establishing the penalties.

Some customer characteristics also affect the request prioritization process. Services that seek to reestablish the energy supply to customers who were previously connected tend to be a priority since there is a higher expectation by the client. A customer who had normal power and a system failure interrupts their supply expects that the company reestablishes the energy as soon as possible. On the other hand, a customer who is requesting a new connection, and does not yet have the power supply, is expected to wait a few days to have a new meter on site. However, the expectation of performing the service for a new connection is higher than the services requested by customers who have standard supply condition, such as for a maintenance of the measuring equipment. Still, customers who pay their bills on time have higher priority than defaulters. A customer interrupted due to a network failure has a higher expectation for a return to the normal state than a customer interrupted due to nonpayment.

These characteristics illustrate the diversity of factors present in the management of the attendance to the requests associated with the energy distribution system. Managing the waiting requests, observing the different priorities inherent to each service, is one of its main challenges.

6.3 Resources

DSOs must have a set of resources to serve the various types of services listed in Sect. 6.2. From the coordination of these resources, it is possible to have teams capable of moving to the places of service in an efficient way to perform the services demanded.

In general, there is an operation center responsible for the planning and monitoring of teams. The concession area of a DSO is usually too large and involves many teams to be managed by one person only. Thus, smaller regions are established in order to enable the planning and monitoring of these teams. Each of these regions must have at least one support facility for the teams. This place is used for storage of materials and vehicles used in network maintenance, training, and administrative activities.

The daily planning of the teams in a region, which defines which requests will be performed and sends this information to the teams, can be performed with different levels of decision support. The supervisor of a region in the operation center can analyze the list of pending requests and select the ones that will be executed by each team, or a computerized decision support system can perform the same task automatically. Although the teams are responsible for serving the requests, the decisions of the operation center have a direct impact on the efficiency of the service. The understanding of the elements involved in the planning decisions is a fundamental aspect for the correct structuring of these decision support systems, and it will be discussed in Sect. 6.4.

From the operational point of view, teams are the main resource of a DSO. They integrate people (electricians), materials, and vehicles. In this section, the emphasis is placed on the characterization of the variations in these three elements and their relationship with the characteristics of the services demanded.

6.3.1 Teams: Types and Characteristics

Correctly serve a request requires the designation of a suitable team for its accomplishment. Requested teams may differ in the amount and qualification of electricians. A single employee can perform services without intervention in the medium-voltage power distribution network. In other cases, at least two employees are required. Services that contain a significant amount of work or that the necessary effort is high will usually require a larger number of people. For example, conducting preventive maintenance involving the exchange of a pole may require teams with five members due to the effort and complexity of the task. However, some complex and riskier tasks do not require a higher number of electricians but require specific experience and training. Live-line working is an example of situations that require more training for electricians.

The designation of a team also requires that it has the necessary materials to perform the request, involving what will be installed, such as cables, meters, and connectors, and the materials needed to perform the task, such as tools, ladders, and protective equipment. These materials must be with the team as it leaves the company's facility and may vary in the amount and volume occupied in the vehicle, as well as the need for protection during transportation to maintain its integrity.

The vehicles used by the teams can be summarized in motorcycles, pick-ups, and trucks. The number of employees is the first factor considered in choosing the vehicle. A service that can be fulfilled by a man can be done with the use of a motorcycle. As the number of employees required increases, the passenger capacity of the vehicle should also increase. The second factor is the type of access to the request location. Urban areas require small size vehicles to facilitate traffic and parking. On the other hand, services in rural areas may require vehicles with high suspension, 4×4 traction, and greater resistance to impact. The type and amount of material to be transported is the third factor that influences the type of vehicle to be used. For example, services on the main circuit breaker that isolates the client from the network do not require ladders, making it possible to use motorcycles, but the services at the service drop involve a ladder, demanding a vehicle with the capability to carry this equipment. Work in places where the ladder does not have access, or that the ladder does not provide the necessary safety to carry out the activity, require the use of a bucket truck. For instance, work in energized networks where it is desired to guarantee higher insulation. The bucket truck allows elevating the electrician to the level of the distribution network without needing support on the pole or cables.

The choice of resources to be used aims to minimize the cost associated with the teams. For this, the set of electricians, materials, and vehicles with the lowest value that meets the requirements of the service is used. As a general rule, the cost increases with the size of the vehicle, number and experience of the electricians, and the quantity and diversity of materials transported. Thus, DSOs use a typical combination of resources to meet the demands, with the objective of executing the services at the lowest possible cost, guaranteeing the necessary quality and safety. This combination may vary between companies, but in general, the relationship between services and resources is established according to the following list.

1. Services without intervention in the medium-voltage power system: teams of one man with a motorcycle

2. Urban regular services (non-emergency): small pick-ups with two men
3. Corrective (emergency) and predictive maintenance services (it does not include the exchange of poles and heavy equipment) or rural regular services: large pick-ups alternating between regular ladders, extendable ladders, and bucket trucks with two men
4. Services with energized networks: bucket trucks with three men
5. Heavy maintenance services (exchange of poles or transformers): truck with five men

In this division, the separation between urban regular services, item 2, and corrective and predictive maintenance services, item 3, depends on the context and presents variations between companies. This variation will be used to illustrate some aspects related to the determination of the services served by each type of team. These two types of teams were composed based on the services requirements. However, they require very similar resources, making economically viable to group their services and to define a multifunctional team to serve them. Nevertheless, grouping the services served by motorcycles, item 1, with urban regular services, item 2, would not be economical, since it would require the most expensive resources (two electricians in a small pick-up) to serve requests that only one electrician could fulfill and would reduce agility in urban displacement. In general, only the use of specific teams for items 2 and 3 or the use of multifunctional teams that serve both groups of services are observed variants.

When specific teams are used, urban regular services are performed by a group of teams and corrective and predictive maintenance services by another group of teams, making management of these two types of demand independent. When using multifunctional teams, the same team can perform both regular services and corrective and predictive maintenance. The use of specific teams presents as advantages: electricians productivity increase by the higher frequency of execution of the same type of service; reduction of the time to master the activity avoiding doubts in the execution of the task; decrease of materials transported by the team, as only the materials, tools, and protective equipment required to perform the tasks that the team attend are required; reduction in the amount of training for each of the teams; and better dimensioning of vehicles per team, requiring a vehicle with lower capacity for urban regular services.

The main disadvantage of this approach is the difficulty of adjusting capacity to variations in emergency demand present in the corrective maintenance services. When there is an increase in this class of demand, the possibility of using the other teams in the fulfillment of waiting requests is not possible or very limited. In this way, some teams tend to become overwhelmed while others may be idle. Another disadvantage is the increase in travel distances, since the more specialized a team is, the less the number of types of services that can be on their route. This limitation in route formation causes two teams to move to the same region to perform different services.

On the other hand, multifunctional teams have the advantage of being able to absorb variations in demand, balancing the resources available between regular

Table 6.1 Situations favorable to specific or multifunctional teams

Situation	Requirement	Team
A high amount of both types of services in urban region	Reducing costs with vehicles, materials and employees. Improvement in the service time through specialist employee	Specific
Low demand for at least one of the service classes	Improving routing and avoiding idleness	Multifunctional
Rural regions or regions made up of small and remote urban centers	Improving routing	Multifunctional
The incidence of high-impact emergency requests	Decreasing emergency waiting time	Multifunctional

services and corrective maintenance as needed. Also, the greater flexibility in the composition of the routes allows reducing the displacements between requests. However, the diversity of services to be carried out tends to reduce the productivity of the teams in the accomplishment of the tasks and to increase the time until the mastery of each one of the types of service. Also, it demands more time dedicated to the training of the teams and raises the number of materials, tools, and protective equipment to be transported.

From the preceding, the use of specific or multifunctional teams is context-dependent, making DSOs decide which strategy to adopt based on the study of their regions. According to Table 6.1, different situations present more affinity with a specific strategy for the team formation. High demand levels concentrated in densely populated regions tend to make travel times smaller, making specific teams perform better in total cost. When one of the classes of services presents low demand, or the traveling distances between orders are high, the construction of efficient routes becomes determinant for the efficiency of the teams, making the performance of the multifunctional teams superior to the specific teams. Finally, when there is a significant incidence of high-impact emergency requests, a shorter response time can be obtained from multifunctional teams, since the team closest to the event will be able to attend.

The decision on the types of teams that the company will use and the respective services served by each one of them is a relevant issue for all electric power distribution companies. However, there is no consensus currently due to the diversity of characteristics found in each region served. Thus, this section presented some service requirements which would influence the definition of these resources.

6.4 Service Operations Planning

The service management in the electricity distribution sector can be analyzed in three horizons: long-, medium-, and short-term. The long-term includes decisions that require more substantial investments and cannot be changed often. Precisely, the company's facilities from which the teams depart are prime examples of this type of investment. Obtaining a property with infrastructure adequate to the needs of this type of installation is not always straightforward, and may require the acquisition of the land and construction of the necessary facilities. Even in cases where it is feasible to lease the infrastructure, observing a horizon of some years to justify the creation of such a structure is required. Thus, the long-term analysis is concerned with determining how the concession area will be divided to allow teams to service the requests efficiently. A facility belonging to each subregion created is responsible for fulfilling its requests, which corresponds to the starting point of the teams working in that subregion. Determining the location of these facilities and their respective coverage region are the primary long-term decisions.

Decisions involving investments that can be revised more frequently than long-term ones are planned in the medium-term. Within this horizon, the central decisions involve the determination of the number of teams and their respective work schedules, with a start and end time of shift, for each of the facilities created from the long-term analysis. These decisions have a certain ease to be reconsidered. However, the hiring and firing costs, the time required to train new teams, the cost of the equipment associated with each team, and the inconveniences caused to employees in shift changes prevent these decisions from being altered often. Thus, long- and medium-term decisions determine all the resources that will be available for the operation at a particular point in time.

The short-term decisions have the purpose of determining how the available resources will be used to service the requests demanded. Within this horizon, it is not possible to significantly change the service's capacity by changing the available resources. Two situations occur in the short-term. Under normal operating conditions, it is necessary to decide which requests will be executed by each team and in which sequence. This decision usually involves regular and emergency requests and can be reviewed at any time throughout the day for the highest efficiency of teams. On the other hand, in extreme emergency situations, in addition to the execution sequence of the orders, one must decide on the displacement of teams between their long-term defined regions. In these cases, the requests involved are only emergencies, with few exceptions where high priority regular requests need to be included.

Next, the aspects that influence decisions in each of the horizons are presented. The understanding of these factors is fundamental to elaborate and evaluate decision support systems in each of the horizons.

6.4.1 Long-Term Planning Horizon

The decision on how to divide the concession area and the location of the facility in each subregion impacts medium- and short-term decisions. On the one hand, these decisions will have an impact on the cost of the teams through the efficiency of the routes generated in the short-term horizon and, on the other hand, the costs associated with the facilities. In a simplified way, the decision in this horizon is to evaluate if the savings generated by reducing the traveling distance of teams and the better response time to emergency orders compensate the cost of a new installation. Higher efficiency in team's route reduces the number of teams required, resulting in operational cost savings, and shorter emergency response times reduce the penalty costs due to non-accomplished deadlines set by regulatory agencies.

Fractioning the concession area limits the work area of a team and the requests it can serve. The area of each subregion will influence team's displacements: the larger the area, the greater will be the displacements. On the other hand, the demand associated with the subregion will determine the number of teams that will benefit from the facility and, therefore, the higher the number of teams required, the greater will be the gain from the reduction of displacements. This trade-off is a challenge when addressing such optimization context.

Predominantly urban areas in large cities tend to justify a more significant number of facilities by generating subregions with a small area and high demand. However, the definition of small subregions tends to be a poor choice in concession areas made up of small and isolated urban centers because of the low demand. In such cases, the small number of teams that will benefit from the facilities tend not to offset the cost of acquiring all infrastructure associated with the installation of a new facility. The low demand of each urban center suggests that greater traveling distances are acceptable as long as they are not frequent.

6.4.2 Medium-Term Planning Horizon

Although in long-term planning horizon there is an estimation of the number of teams that will be active in each subregion, in the medium-term planning horizon this decision is reviewed along with the definition of each team's shift work schedule. These decisions are reviewed more often than long-term decisions, allowing to adjust the service capacity to changes in demand. Also, each of the subregions defined in the long-term usually have autonomy to plan their teams' schedule.

The decision on the number of teams is strongly related to the company's decision on the response time to the requests. With more staff, the cost of the operation increases, but the response time to customer requests and network repairs in emergencies is lower, thus leading to a reduction in penalties. On the other hand, the cost of the operation decreases with the reduction of the number of teams. However, the response time tends to increase, generating a backlog of both regular and emergency requests. The determination of the backlog due to a certain number of teams and the consequent cost of delay in request completion depends on the variability of the demand and the start and end of teams' workday hours.

Both regular and emergency demand have a significant degree of uncertainty, regardless of extreme demand cases. The higher the demand variability, the greater will be the number of requests that experience high waiting time. It means that the number of teams should consider the uncertainty in the evaluation of the costs associated with the delay in requests fulfillment. For example, suppose that the demand for services is 8 h per day, without variability. In this case, just one team working 8 h per day would be enough to meet the demand, and there would not be a backlog at the end of every planning day. However, if there is significant variability, backlog formation occurs even with a balance between capacity and demand. For example, if the demand on the first day was 16 h, on the second day zero hours, on the third day 16 h, on the fourth day zero hours, and so on, keeping the average capacity at 8 h per day will result on 24 h of waiting time for some requests. Since the penalty costs are associated with the time that requests remain in backlog until its competition, it is necessary to consider the effects of demand variability when defining the most appropriated number of teams. Therefore, one must consider that part of the variability can be absorbed with the use of overtime, always respecting the related labor legislation.

The decision on the start and end times of the teams' shift work schedule has an impact mainly on the emergency backlog. Regular requests have longer completion times, around few days, and therefore, the backlog is usually evaluated only once a day. The most common restriction related to this type of request is related to the execution: they will only be executed within business hours and business days. In this way, the teams' scheduling does not have as much impact on the regular backlog as long as it is within the time window considered as business hours. Emergency requests have shorter completion times, measured in hours, and the corresponding backlog is permanently monitored to avoid peaks throughout the day. Hence, teams' scheduling has a direct influence on the backlog level, since the lack of team's availability at times of high demand leads to a quick increase in the delay, what may conduct to penalty costs. In order to get shorter turnaround time and avoid these costs, more teams should be working at peak demand hours.

Another characteristic to be considered in defining the number of teams is the travel time. The time spent traveling to the request's position reduces the time that the team has available to meet the demands. It should be estimated according to the coverage area of the subregion under analysis and is related to the long-term decision. Larger areas will have a longer average displacement time than smaller ones, having the teams' efficiency more impaired.

Defining teams' scheduling must address another characteristic of the distribution sector: the availability of teams for 24 h a day. On-call shifts can provide service availability at low demand periods, for instance between midnight and 6 o'clock. In these cases, a set of technicians remain on duty for eventually attending requests. The corresponding response times related to teams enrolled in this type of shift involve the preparation time when it is activated, not required in cases when the team is already engaged.

6.4.3 Short-Term Planning Horizon

The short-term horizon corresponds to the planning on the use of defined resources in the long- and medium-term. Under normal conditions, the demand to be met by the teams of a particular facility is restricted to its coverage area. Thus, the decision to be made is to determine which requests will be served by which team and in what sequence. However, such planning can be done in advance only for regular requests, as these have more extended deadlines and can be planned on a daily basis. Thus, when deciding on the teams' routes, all available regular requests are known, improving the quality of the solution. On the other hand, emergency requests require shorter completion times, requiring more reactive planning. When using teams that meet both regular and emergency requests, it is possible to obtain reduced response time in certain situations, but this establishes some interdependence between such decisions.

The planning of the initial routes involves only regular requests and considers the minimization of the teams' displacements since these orders are spatially distributed. Figure 6.1 illustrates this situation where teams A and B are available, the first one having a 9-request route and the second has a route with eight requests. In this hypothetical scenario, four requests (yellow circles) cannot be included in the planned route because they would result in an extrapolation of the workday hours of teams A and B, conducting the left requests to remain in the backlog to further competition on the next day schedule.

However, this is not the primary feature to consider at this planning stage. Managing the backlog to meet all regular requests within the defined deadlines is the main objective to be achieved, what become even more challenging with a given working capacity, usually insufficient to accomplish all requests in backlog. Therefore, it is important to assign priorities to requests, dividing the backlog into subsets with different degrees of urgency. More specifically, other factors beyond its deadline influence the priority of a request, such as its type and the type of customer affected. Therefore, the priority should not only be used to determine which orders

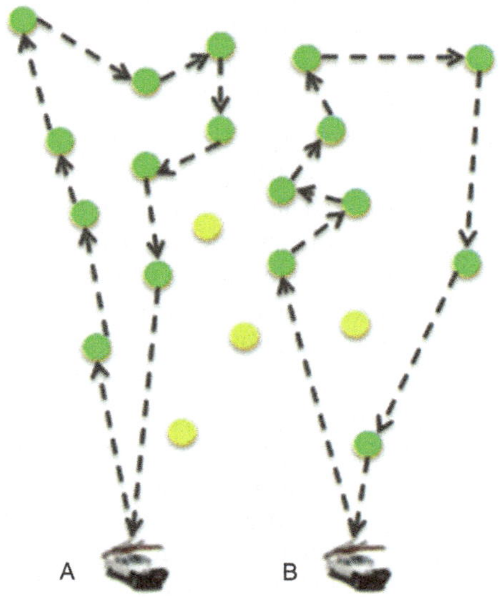

Fig. 6.1 Two teams and their hypothetical routes

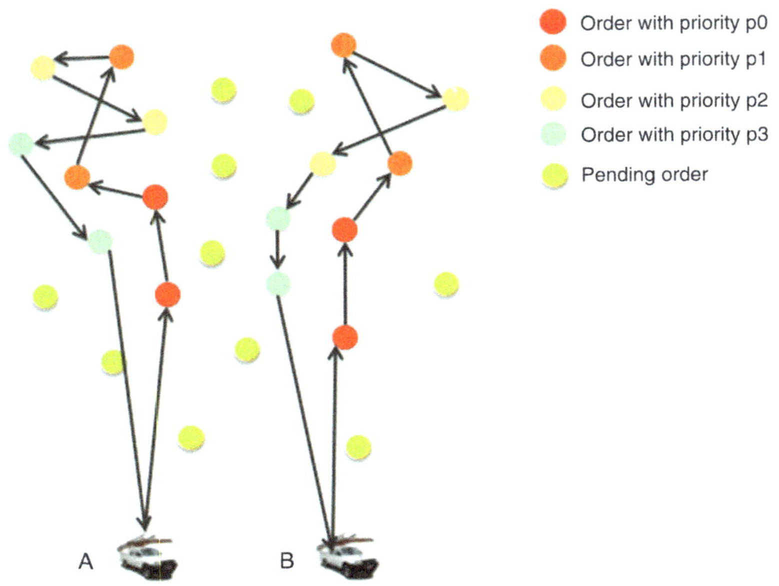

Fig. 6.2 Two teams, their hypothetical routes with order priorities defined

will be dispatched to the teams and which ones will remain in the backlog for the next day but should also influence the sequence of the orders within a route. Due to possible arrival of emergency requests, it is necessary to ensure that some requests are more likely to be met than others, depending on the nature of services and deadlines. In this way, the routes must strictly obey the priority precedence defined a priori, as shown in Fig. 6.2. Routes are substantially modified to reflect the sequence p0, p1, p2, p3, which makes the process less efficient and causes even more requests to be included in the backlog (yellow circles).

Also, the entry of emergency requests on the route significantly affects the time spent in displacement, as shown in Fig. 6.3 in contrast with Fig. 6.1. Considering the existence of emergencies when planning the initial route of teams is one way of minimizing such an adverse effect. For instance, spread teams in areas where there is a historically high incidence of emergencies may be a way of reducing response time. It prevents most teams from being away from these locations or that all teams may be concentrated in a particular region.

It is worth noting that the approach assumed to address emergency occurrences follows the planning logic to be applied typically in a day planning basis, with few variations on the horizon of 1 week. This assumption is justified because what is expected is to estimate an average daily level of emergency demand and include this amount in the planning of the teams so that it is possible to construct the routes respecting a "reserve" of capacity to take requests from this nature. Atypical situations related to extreme climatic events, precisely because of the unpredictability and the particular and individualized treatment condition, cannot

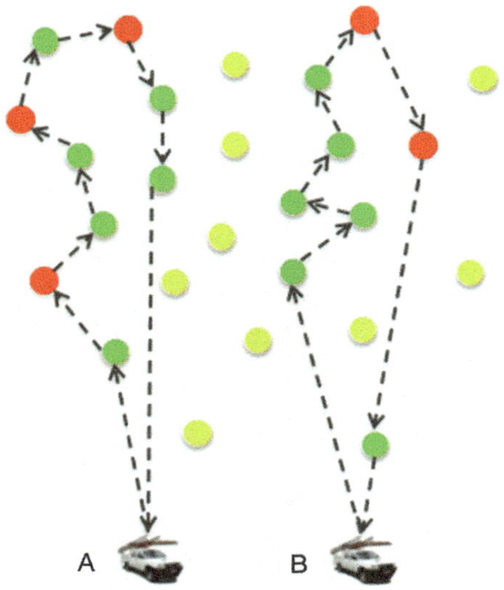

Fig. 6.3 Two teams and their hypothetical routes, when emergency orders are assumed

be included in this approach, requiring a specific procedure. It involves not only dedicating some teams to the emergency requests from the associated climatic event, but also the displacements of teams between the facilities.

Another approach to reducing the complexity associated with planning several routes refers to the subdivision of the area into smaller parts so that each one is the set of requests (demand) to receive service of one or several teams. Restricting teams' coverage area is justified in some subregions due to the peculiar characteristics associated with traffic or even due to the constructive form of the network that may include certain type of customers. Figure 6.4 illustrates this subdivision.

From the subdivision of the area, as shown in Fig. 6.4, the assignment of teams to the considered subareas is carried out, as shown in Fig. 6.5. In this example, it is possible to note that teams are restricted to the subareas and that certain requests are not included on the routes defined (green circles) because they cause violation on the teams' working day hours. After dividing the area and assigning teams to them, it is possible to verify the balance between capacity (number of teams' working hours) and demand (number of hours required to carry out all requests). In this context, a procedure is used to define the requests' priorities used in the route construction. With this approach, it is possible to address the complexity of the associated problem through the elaboration of routes restricted to the defined subareas. Although the geographic area is divided to separate the regions that involve the teams' operation, they have the same starting point: the geographical location associated to the facility planned to the subregion defined in the long-term horizon.

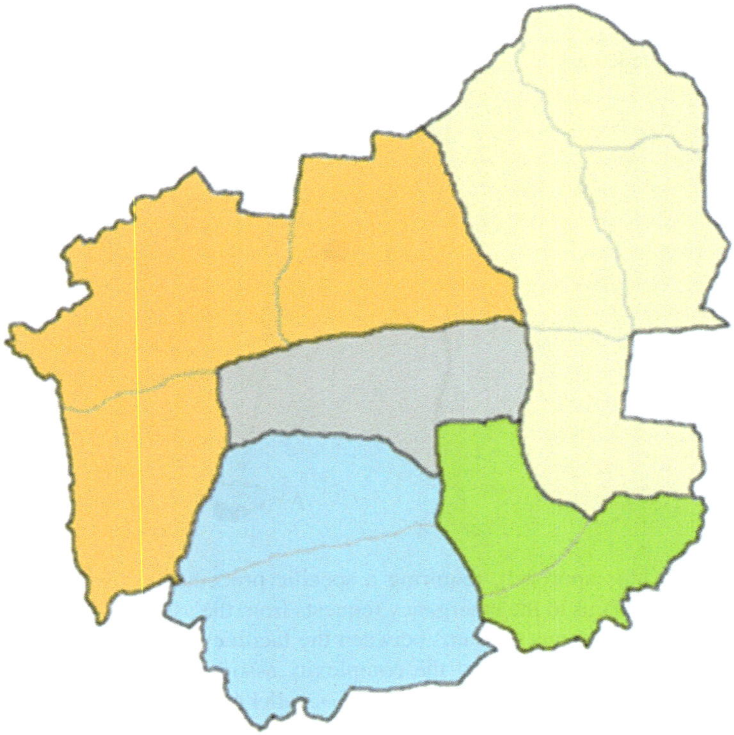

Fig. 6.4 A possible division of the service area

6.5 Conclusions

The management of services related to the energy distribution system has a direct impact on the population. Also, the diversity of services provided and the influence of regulatory agencies makes the service management a challenge for DSOs. The resources planning aims to guarantee the desired level of service at the lowest cost. Understanding the characteristics of the demand and the resources used in the daily routine is fundamental to the design of the operation. In essence, the medium- and long-term decisions seek to ensure the efficiency of the teams in the short-term. By understanding the challenges faced by the operations center, it is possible to comprehend how the medium- and long-term decisions can afford better conditions to provide an efficient service.

The most prominent challenge in the short-term involves integrating backlog management with team routing, which is highly dependent on the previous decisions. In the long-term, the decision on how to divide the concession region and the location of the facilities has a direct impact on the routes of the teams in the short-term. In the medium-term, the planned number of teams determines whether the

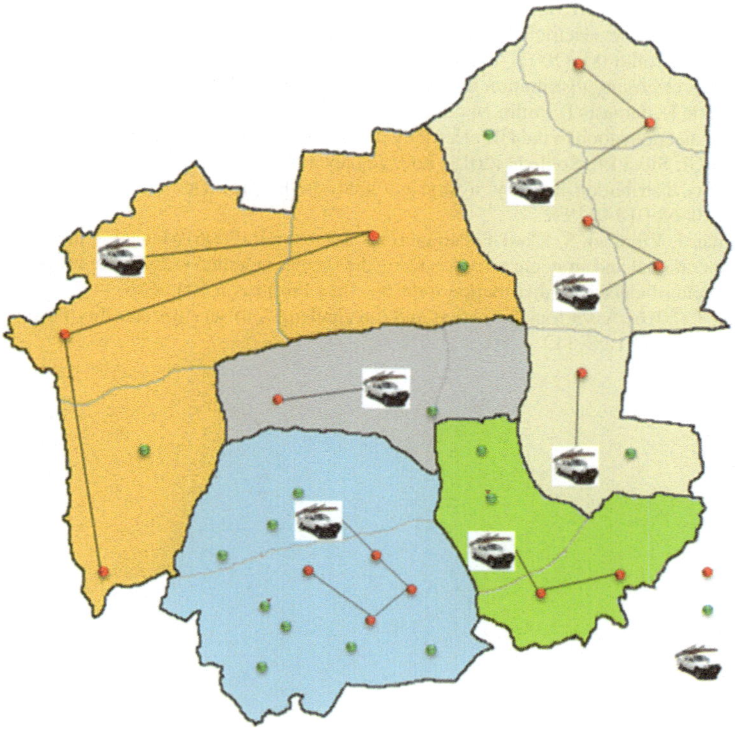

Fig. 6.5 The service area divided, and the corresponding teams associated

service will be capable of meeting the response times defined by the company. Also, the definition of the team's shift work schedule has a direct impact on the backlog of emergency requests. From these relationships, the integration between these planning horizons is required, as well as the development of computational decision support systems able to consider the various factors inherent to each of these decisions.

References

1. Willis HL (2004) Power distribution planning reference book, 2nd edn. CRC Press, New York
2. Gonen T (2016) Electric power distribution engineering, 3rd edn. CRC Press, New York
3. Perrier N, Agard B, Baptiste P, Frayret J, Langevin A, Pellerin R, Riopel D, Trépanier M (2013) A survey of models and algorithms for emergency response logistics in electric distribution systems. Part II: contingency planning level. Comput Oper Res 40:1907–1922
4. El Hage FS, Rufín C (2016) Context analysis for a new regulatory model for electric utilities in Brazil. Energy Policy 97:145–154

5. Filippini M, Greene W, Masiero G (2018) Persistent and transient productive inefficiency in a regulated industry: electricity distribution. Energy Econ 69:325–334
6. Jamasb T, Pollitt M (2003) International benchmarking and regulation: an application to European electricity distribution utilities. Energy Policy 31:1609–1622
7. Giannakis D, Jamasb T, Pollitt M (2005) Benchmarking and incentive regulation of quality of service: an application to the UK electricity distribution networks. Energy Policy 33:2256–2271
8. Pereira GI, Silva PP, Soule D (2018) Policy-adaptation for a smarter and more sustainable EU electricity distribution industry: a foresight analysis. Environ Dev Sustain. https://doi.org/10.1007/s10668-018-0119-x
9. Bergman J, Viljainen S, Kässi T, Partanen J, Laaksonen P (2006) Managing the exploration of new operational and strategic activities using the scenario method—assessing future capabilities in the field of electricity distribution industry. Int J Prod Econ 104:46–61
10. Helms T (2016) Asset transformation and the challenges to servitize a utility business model. Energy Policy 91:98–112

Chapter 7
The Dispatch Problems in Power Distribution Systems

M. Schmitz, C. H. Barriquello, and Vinícius Jacques Garcia

7.1 Introduction

A dispatch problem is a kind of optimization problem which occurs in several fields as engineering, economics, logistics, and so on. In a general formulation, in a dispatch problem, the goal is to dispatch a set (or subset) of processors (e.g., vehicles, crews, providers, machines) ($P = \{P_1, P_2, \dots, P_m\}$ of m processors) to a set (or subset) of tasks (e.g., jobs, customers, service orders, places, loads) ($T = \{T_1, T_2, \dots, T_n\}$ of n tasks), while maximizing (or minimizing) an objective (or multiobjective) function subject to a set of given constraints (e.g., capacities, deadlines).

Some examples of dispatching problems are:

- Truck Dispatching Problem [1]
- Train Dispatching Problem [2]
- Elevator Dispatching Problem [3]
- Ambulance Dispatch Problem [4]

In this chapter, we focus on the dispatching problems occurring in electric distribution power systems. Examples of such problems which are presented along

M. Schmitz (✉) · C. H. Barriquello · V. J. Garcia
Technology Center, Federal University of Santa Maria, Santa Maria, Rio Grande do Sul, Brazil
e-mail: barriquello@mail.ufsm.br; viniciusjg@ufsm.br

D. P. Bernardon, V. J. Garcia (eds.), *Smart Operation for Power Distribution Systems*, https://doi.org/10.1007/978-3-319-93922-3_7

the chapter include the dispatch of repair crews for servicing emergency orders in the distribution power system and the dispatch of distributed generators with minimum generation cost, also known as economic dispatch.

The chapter is organized as follows: in Sect. 7.1 we present an overview of the dispatch problem. Then, in Sect. 7.2 we study the economic dispatch problem, including the centralized and the distributed versions of the problem. Next, in Sect. 7.3, we study the service crews dispatch problem, which is the problem of dispatching crews for services in the distribution power system. Finally, we conclude this chapter in Sect. 7.4, remarking the importance of the research of new formulations and algorithms of the dispatch problems in the smart power distribution systems.

7.2 An Overview of a Dispatch Problem

Depending on the specifics of the dispatch problem formulation, a general dispatch problem can be broken into one or more classes of problems (or subproblems). This view of the problem allows for adapting the problem formulation and the algorithms according to specific scenarios of application. Therefore, in this chapter, we view a dispatch problem as formed by one or more of the three classes of optimization problems: assignment, routing, and scheduling. An illustration of the proposed view is given in Fig. 7.1.

Indeed, according to a given scenario, some aspects of the problem can be more relevant or, on the contrary, may not even be necessary. Therefore, correctly identifying the class(es) of dispatch problem is the first step toward solving it.

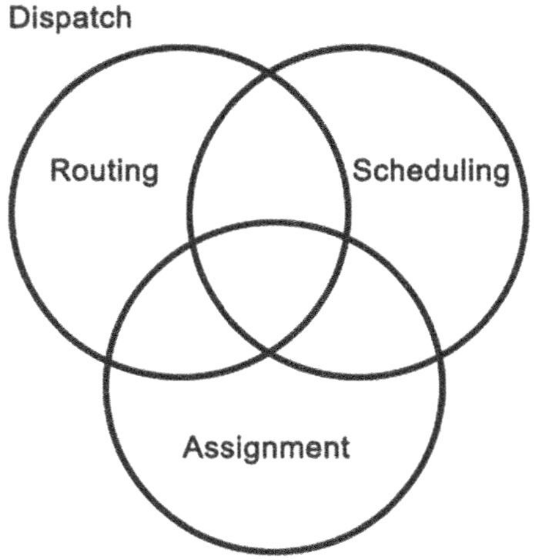

Fig. 7.1 A general view of a dispatch problem

7.2.1 Assignment

An assignment problem deals with the question of how to assign n *tasks* to n *processors* in the best possible way. The problem consists of two components: the *assignment* as underlying combinatorial structure and an objective function modeling the "best way" [5].

Therefore, an assignment is a bijective mapping between two finite sets: P of processors and T of tasks. In this way, an assignment can be represented as a *permutation* ρ, such that element *1* is mapped to $\rho(1)$, element 2 mapped to $\rho(2)$, ..., n is mapped $\rho(n)$.

Indeed, in general one can have several possible assignments between the two sets P and T. Thus, it is interesting to pose the assignment problem as an optimization problem, asking the question: what is best suited assignment among all possible assignments S_n for the problem under consideration? In this case, we must have an objective function representing our goal for the problem and an $n \times n$ cost matrix $C = (c_{ij})$ representing the cost of assigning i to j.

7.2.2 Routing

As defined in [6, 7], given a set of transportation requests and a crew of vehicles, the vehicle routing problem (VRP) consists in determining a set of routes to cover all transportation requests, starting and ending at its own depot, with the given crews at a minimum cost, defining which vehicle attend which request in which sequence such that all requests are attended. Also, the Capacitated VRP ensures that the capacity of the vehicles is not exceeded. The VRP has numerous variations, which introduce characteristics of real life, like the VRP with variation in capacities, which results in the heterogeneous fleet VRP, or the popular extension VRP with time windows, which assumes that the requests must be attended in a certain time interval, which varies from task to task.

7.2.3 Scheduling

Scheduling problems can be generally understood as optimization problems for resources allocation over time, in order to cover a set of tasks belonging to a process [8, 9].

Scheduling problems are usually characterized by three sets: set $T = \{T_1, T_2, \ldots, T_n\}$ of n *tasks* (commonly referred to as *jobs*), set $P = \{P_1, P_2, .., P_m\}$ of m *processors* (*crews*, *machines*), and set $R = \{R_1, R_2, \ldots, R_s\}$ of s types of *additional resources R*. Scheduling, generally speaking, means to assign processors from P and (possibly) resources from R to tasks from T in order to complete all tasks under the imposed constraints [8, 10].

7.2.4 The Dispatch Problem

As seen so far, the dispatch problem can be approached as a scheduling, assignment, or routing problem. In addition to these, the dispatch can be approached in a hybrid way as a combination of these problems, obtaining the variations presented below and illustrated in Fig. 7.1:

- Routing and Assignment:

  The routing and assignment can be applied as presented by [11], where the author propose a methodology to *assign* to each customer a time window from a set of candidate time windows, before demand is known. In a second stage, when demand is revealed, vehicle *routes* satisfying vehicle capacity and the assigned time windows are constructed.

  [12] presents a technique to cluster branches into groups belonging to each depot and to produce the routes for each depot. The assignment problem (AP) and vehicle problem with time window (VRPTW) is applied in cash distribution of a bank.
- Routing and scheduling:

  A vehicle routing-scheduling problem is characterized as a routing problem complicated by several time windows and inter-arrival time constraints at each customer point. The approach can be applied, for example, for the machine or production scheduling, as presented by [13, 14].

  In [13], the author proposes a nonlinear mathematical model to consider production *scheduling* and vehicle *routing* with time windows for perishable food products in the same framework. The perishable goods will deteriorate once they were produced. The author elaborates a heuristic for the vehicle routing with time windows.

  In [14], the author first addresses scheduling of a set of jobs on parallel machines with machine-dependent ready times and then solves the delivery of completed jobs with a fleet of vehicles which may differ in their loading capacities and ready times. In [15] a survey is presented.
- Assignment and scheduling:

  The assignment and scheduling problem can be applied as follows: Given a set of tasks, determine which tasks are to be *assigned* to each processor, find a *schedule* for processing the assigned tasks at each processor.

  In [16], the author studies the order assignment and scheduling of the time-sensitive products manufacture. The processing time and cost of an order are dependent on the plant to which it is assigned. It is determined which orders are to be assigned to each plant. Also, it is found a schedule for processing the assigned orders at each plant, and a schedule for shipping the completed orders from each plant to the distribution center, such that a certain performance measure is optimized.

  In [17], the authors study a due-date assignment with the option of scheduling a maintenance activity. Jobs scheduled before (after) the due-date are penalized according to their earliness (tardiness) value.

- Assignment and routing and scheduling:

  As said before, the assignment, routing, and scheduling problems can be applied in an integrated way, as presented by [18] where the author formulated a problem as a combination of assignment, scheduling, and routing problems. The author considers a real-life case study that determines the minimum number of sellers required to attend a set of customers located in a certain region taking into account the weekly schedule plan of the visits, as well as the optimal route. The problem is divided into three steps, in step I, a set of cost-effective feasible clusters of customers assigned to a seller. Steps II and III are then used to solve the problem of a weekly program for visiting the customers as well as to determine the route plan using MILP formulation.

In the following sections, we introduce the reader to two instances of dispatch problems that power system operators may face: (1) the economic dispatch problem (EDP) and (2) the service dispatch problem (SDP).

7.3 Economic Dispatch Problem

The economic dispatch problem (EDP), also known as optimal dispatch, is one of the well-known and fundamental problems in electric power systems. In short, it is the problem of minimizing the cost of the generated power subject to the constraints of load demand, capacities of the generation, transmission and operation facilities of the system and the required reliability.

Although first works on this problem date back to the beginning of the twentieth century [19, 20], it has received renewed and increased attention in the last decade within the smart grid agenda, where new elements must be taken into account (e.g., renewable energy sources—RESs, distributed generation units and price-responsive loads, to name a few) [21, 22]. Since then, new approaches for the problem of supply and demand balance have been proposed for including several aspects of the smart grid. Therefore, not only the economic aspect is considered, but also the environment impact [23], the reliability of the system [24], the distributed and real-time information available [25, 26], and the controllability of the loads [27]. In the sequel, we revisit the classical optimal dynamic economic dispatch problem formulation and provide an overview of the new formulations of the dispatch problem targeted for the smart grid and for the dispatching of the distributed generation units (GUs) installed at the distribution system.

7.3.1 Optimal Dynamic Economic Dispatch

The classical optimal dynamic economic dispatch problem (ODD) can be posed as an optimization problem where the goal is to minimize the production cost subject to a set of constraints [28]. In this formulation, the forecast load demand is assumed to be available over the entire dispatch period $[0, NT]$ and it is given at discrete-time intervals $[kT, (k + 1)T]$, $k = 0, 1, \ldots, N - 1$, where T is the sampling period and N is the number of sampling periods. Usually a fixed sampling period is considered. Moreover, the following definitions are given:

n: Number of committed generation units
L: Number of transmission lines
P_i^t: The generation of unit i during the time interval $[(t - 1)T, tT]$
$C_i(P_i^t)$: The generation cost for unit i to produce P_i^t
D^t: The demand at time t (i.e., the tth time interval)
U_i^t: The control variable for the ramp rate of the unit i at time t
UR_i and DR_i: The maximum ramp up/down rates for unit i
$P^{\min}{}_i$ and $P^{\max}{}_i$: The minimum and maximum capacity of unit i, respectively
S_i^t: The spinning reserve contribution of unit i during the time interval t
SR^t: The system spinning reserve requirement for interval t
$F^t{}_l$: The active power flow through transmission line l during the interval t
$F^{\max}{}_l$: The upper limit on the active power flow along line l

Now let us start by the first formulation (DED), which is given by the objective function (Eq. 7.1) and the constraints (Eqs. 7.2–7.8):

$$\min C = \sum_{t=1}^{N} \sum_{i=1}^{n} C_i(P_i^t) \tag{7.1}$$

Subject to Eqs. (7.2–7.8) for $t = 1, 2, \ldots, N - 1$, $i = 1, 2, \ldots, n$ and $l = 1, 2, \ldots, L$.

$$\sum\nolimits_{i=1}^{n} P_i^t = D^t + P_{\text{loss}}^t \tag{7.2}$$

$$-DR_i \cdot T \le P_i^{t+1} - P_i^t \le -UR_i \cdot T \tag{7.3}$$

$$P_i^t + S_i^t \le P_i^{\max} \tag{7.4}$$

$$P_i^{\min} \le P_i^t \tag{7.5}$$

$$0 \le S_i^t \le UR_i.T \tag{7.6}$$

$$\sum\nolimits_{i=1}^{n} S_i^t \ge SR^t \tag{7.7}$$

$$-F_l^{\max} \le F_i^t \le F_l^{\max} \tag{7.8}$$

The constraints are the load-generation balance (Eq. 7.2), the ramp rate limits (Eq. 7.3), the maximum and minimum capacities (Eqs. 7.4–7.5), the maximum-ramp spinning reserve contribution (Eq. 7.6), the system spinning reserve requirement (Eq. 7.7), and the line flow limits (Eq. 7.8).

The transmission line losses and the line flow are, respectively, given by Eqs. (7.9) and (7.10).

$$P_{\text{loss}}^t = \sum_{j=1}^{n} \sum_{i=1}^{n} P_i^t B_{ij} P_j^t + \sum_{i=1}^{n} B_{i0} P_i^t + B_{00} \tag{7.9}$$

$$F_l^t = \sum_{i=1}^{n} \rho_{l,i} P_i^t \tag{7.10}$$

Where B_{ij} is the ijth element of the loss coefficient square matrix, B_{i0} is the ith element of the loss coefficient vector, B_{00} is the loss coefficient constant, and $\rho_{l,\ i}$ is the generalized generation distribution factors.

Then, the fuel cost functions $C_i(.)$ is based from the fuel consumption function that can be measured. Commonly used cost functions are the smooth quadratic cost function (Eq. 7.11) and the nonsmooth cost function that considers the valve-point effect (Eq. 7.12), where a_i, b_i, and c_i are positive constants, and d_i and e_i are the coefficients of generator i reflecting valve-point effects.

$$C_i\left(P_i^t\right) = a_i + b_i P_i^t + c_i\left(P_i^t\right)^2 \tag{7.11}$$

$$C_i\left(P_i^t\right) = a_i + b_i P_i^t + c_i\left(P_i^t\right)^2 + \left|d_i \sin\left(e_i\left(P_i^{\min} - P_i^t\right)\right)\right| \tag{7.12}$$

As formulated, the economic dispatch problem can be identified as a scheduling problem constrained by the routing component (due to network losses and the line flow capacities). However, the assignment of the generators to loads is not relevant due to the nature of the problem (i.e., only the total power needs to meet the total demand).

Indeed, the scheduling aspect of the problem tends to be exacerbated in future power systems, where the presence of the intermittent power generators (e.g., solar and wind power) will be massive at the distribution side. This brings a challenge to the system operators, as the economic dispatch problem becomes distributed and real-time [29].

7.3.2 Economic Dispatch for Distributed Generators

The classical economic dispatch formulation was developed for a power grid where generator outputs could be controlled as needed to continuously deliver power to the loads. However, these generators were mostly based on fossil fuel and emitters of gaseous pollutants such as SO_2, NO_x, CO, and CO_2, which affect human health and the environment. On the other hand, the power grid has increasingly received more generators based on renewable sources, such as wind and solar, and which are fundamentally different than the fossil-fuel-based generators.

The renewable generators are intermittent, and cannot be operated as conventional generators. Moreover, due to the possibility of deploying renewable generators in a small scale at the distribution network or at consumer side, the number of installed distributed generators (DG) is expected to increase. More and more distributed energy resources (DERs), such as distributed generation units and distributed energy storage devices, will be integrated to the grid along with communication networks as the grid evolves toward to the smart grid. Therefore, as the number of DG units grows, economically dispatching those generators in a centralized manner becomes more challenging due to the required computation and communication burden.

Given such large-scale and distributed nature of the problem, it becomes attractive to adopt a distributed approach to solve the economic dispatch problem. Therefore, recently several works have been published with possible algorithms to solve the problem of the real-time and distributed economic dispatch. These algorithms must deal with the distributed nature of the problem and with the communication among the generators. Some of the proposed approaches include:

- Consensus algorithms [26, 30, 31]

  Consensus algorithms have been proposed to solve the economic dispatch problem in a distributed way. This approach is based on the fact that in the optimal dispatch, the incremental costs of all generators are equal. However, the consensus algorithms need to be extended in order to satisfy the constraint that the total generation in the system needs to be equal to the total load.
- Game theoretic approach [26]

  Game theory can be used as tool for solving distributed optimization problems in a muti-agent system, where agents are self-optimizing and converge to an equilibrium. In this approach, game theory allows for the decomposition of the optimization problem (game design) into local objective functions and the specific local decision rules (distributed learning algorithms) [32]. Therefore, the collective behavior of the agents can converge to the desired equilibrium based solely on local interactions and local decisions taken by the agents. This approach has been used in [26] to design the game and the local control laws to solve the real-time economic dispatch problem in a distributed way.
- Virtual power plant (VPP) [33, 34]

  In a virtual power plant approach, small generation units and storages at different locations are grouped by a central entity, called energy management system (EMS), in order to reach necessary minimum capacities for market participation. Moreover, the EMS can optimally dispatch the DG units working as central coordinator. However, if the DG units are deployed at dispersed locations, the cost of gathering the required information may be prohibitive, and thus a decentralized dispatch mechanism is still required.

7.4 Service Dispatch Problem

The service dispatch problem basically consists of assigning and scheduling vehicles (*P*) (maintenance and repair crew) to service orders (*T*), thus forming routes, in a way that all (or some) orders are attended at a minimum cost. These routes have their start and end points in a depot. Usually, the objective is to minimize the total route cost, namely the sum of costs associated with the fleet size and the costs of completing the delivery tours. Reducing the dispatch time is also an objective, mainly when dealing with emergency orders. There are two classical constraints on the dispatch problem in the power system. Each order must be executed by at most one vehicle at a time and each vehicle is able to process at most one order at a time.

7.4.1 Dispatch Problem Variants and Formulations

The vehicle routing problem consists of designing the routes for the vehicles, assigning each order to a vehicle, and sequencing the locations to be visited. To a better understanding, we will give now a short definition of each one.

7.4.1.1 Scheduling

The scheduling problem consists of deciding the time at which each vehicle starts the service at each location [6].

On the electric power distribution systems vehicle and service orders scheduling problems, the optimality criteria can be the *completion time* of order j, C_j, $\forall j \in T$; the *flow time* of order j, $F_j = C_j - r_j, \forall j \in T$(the sum of waiting and processing times. r_j is the release time of order $j \in T$); the earliness of order j, $L_j = C_j - d_j$, $\forall j \in T$; the *tardiness* of order j, $D_j = \max\{C_j - d_j, 0\}$ (where d_j is the due date of task j) and the *unit penalty* $U_j = 0$ if $C_j \leq d_j$, 1 otherwise. The most commonly chosen involves minimization of $f_{\max} \in \{C_{\max}, L_{\max}\}$, minimizing the average time each customer is without power [10, 35].

A generic single-processor scheduling problem can be formulated as an integer linear programming (ILP) by introducing a binary variable x_{ij}, which will assume the value 1 only if the order i precede order j on the schedule, $(i, j = 1, \ldots, |T|)$:

$$\min C_{\max} \tag{7.13}$$

s.t.

$$\sum_{j \in T} x_{ij} = 1, \forall i \in T \tag{7.14}$$

$$\sum_{j \in T} x_{ji} = 1, \forall i \in T \tag{7.15}$$

$$C_j \geq C_i + \mathrm{ST}_j + M(x_{ij} - 1), \forall i, j \in T \backslash T_o \tag{7.16}$$

$$C_{\max} \geq C_j, \forall j \in T \tag{7.17}$$

$$C_k = 0, \forall k \in T_o \subseteq T \tag{7.18}$$

$$x_{ij} \in \{0, 1\}, \forall i, j \in T \tag{7.19}$$

where ST is the service time.

$$\phi = (4, 1, 3, 2)$$

$$X_\phi = \begin{pmatrix} 0 & 1 & 0 & 0 \\ 0 & 0 & 0 & 1 \\ 0 & 0 & 1 & 0 \\ 1 & 0 & 0 & 0 \end{pmatrix}$$

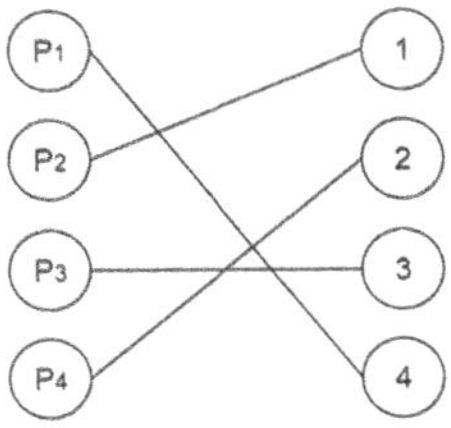

Fig. 7.2 Different representations of assignments

7.4.1.2 Assignment

Finally, the crew assignment problem consists of assigning a set of crews to *resource depots*, so as to satisfy the demand for repair tasks while minimizing delays and costs [36–38].

An *assignment* is a bijective mapping of a finite set into itself, i.e., a *permutation*. As defined in [5], every permutation ϕ of the set $P = \{1, \ldots, n\}$ corresponds in a unique way to a $n \times n$ *permutation matrix* $X_\phi = (x_{ij})$(Fig. 7.2) with:

$$x_{ij} = \begin{cases} 1, & \text{if } j = \phi(i) \\ 0, & \text{otherwise} \end{cases}$$

The set of all assignments (permutations) of n items will be denoted by Sn and has $n!$ elements. We can describe this set by the following equations called *assignment constraints*.

$$\min \sum_{i \in T} \sum_{j \in T} c_{ij} x_{ij} \tag{7.20}$$

s.t.

$$\sum_{i=1}^{n} x_{ij} = 1, \forall j = 1, \ldots, n \tag{7.21}$$

$$\sum_{i=1}^{n} x_{ji} = 1, \forall j = 1, \ldots, n \tag{7.22}$$

$$x_{ij} \in \{0, 1\}, \forall i, j = 1, \ldots, n \tag{7.23}$$

The Eq. (7.20) says that every row of a permutation matrix sums to 1. The Eqs. (7.21) and (7.22) says that every column of a permutation matrix has a sum of 1. Finally, Eq. (7.23) define that a permutation matrix has only the entries 0 and 1.

7.4.1.3 Routing

The vehicle routing dispatch problem can be defined as [6]: Given a set of service orders and a crew of vehicles, the dispatch center must determine a set of vehicles

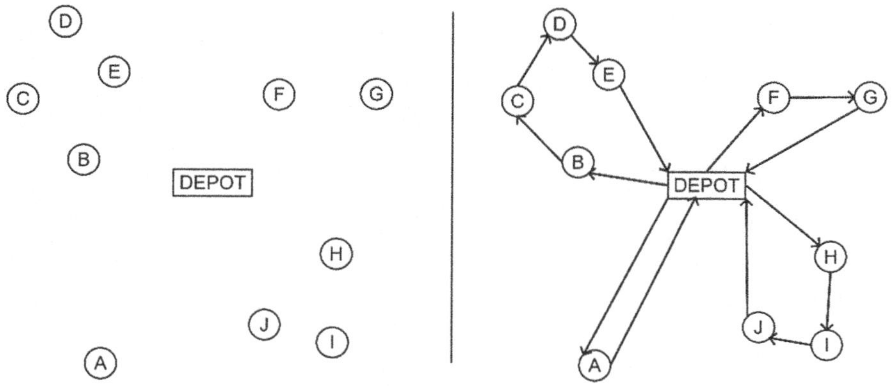

Fig. 7.3 Example of the two-phase *cluster-first route-second* procedure in vehicle routing

routes to attend all (or some) service orders at a minimum cost; defining which vehicle handles which order in which sequence.

On the electric power distribution systems vehicle routing problem, the optimality criteria can include *completion time* C_j; the *lateness* L_j; *unit penalty* U_j; overall routing cost c_{ijp}; *crew size* minimization $\min(p)$; and minimizing the length of the longest route.

Before the routing step, clusters of service orders are defined, determining the orders compatible with vehicle capacity, for then applying the best vehicle routing strategy for each cluster. This strategy is the well-known *cluster-first route-second* method, like the sweep heuristic [39]. The opposite approach can also be applied, known as *route-first cluster-second.* We will here consider that a *cluster-first route-second* method (Fig. 7.3) was applied, obtaining clusters of service orders (subset of T), each one attended by a vehicle (p).

From the consideration of the order set partition, the following problem address the corresponding routing of orders assuming that there will be a single vehicle (p) previously assigned to attend this set (T). This formulation thus answers which will be the visiting order sequence, getting it closest to the Traveling Salesman Problem formulation [40] together with the MTZ formulation to avoid subtours [41, 42]:

$$\min \sum_{i \in T} \sum_{j \in T} c_{ij} x_{ij} \tag{7.24}$$

s.t.

$$\sum_{j \in S} x_{ij} = 1, \forall i \in T \tag{7.25}$$

$$\sum_{j \in T} x_{ji} = 1, \forall i \in T \tag{7.26}$$

$$u_i - u_j + 1 \leq (|T| - 1)(1 - x_{ij}), \quad \forall i,j \neq 1 \tag{7.27}$$

$$2 \leq u_i \leq |T| \quad (7.28)$$
$$u_1 = 1 \quad (7.29)$$
$$x_{ij} \in \{0, 1\}, \forall i, j \in T \quad (7.30)$$
$$u_i \in \mathbb{N}, \forall i \in T \quad (7.31)$$

where u_j is the integer variable that avoid subroutes and c_{ij} is the travel cost of a vehicle moving from i to j.

Given the above definition of different dispatch variants (scheduling, routing and assignment) in electric power distribution systems, we will from now on focus on vehicle routing dispatch problem.

7.4.2 *The Offline and Online Dispatch Problem in Electric Power Systems*

The vehicle dispatching problems in the electric power distribution systems is composed of two fundamental types of service orders, the customer, the maintenance and the emergency ones. The customers—from power outages, new branch connections—and maintenance ones—from inspection and maintenance of equipment—have all the input parameters known a priori, in the planning phase. The demand for emergency services comes from interruptions in energy supply—from climate events, equipment failures, and human—with unforeseen time and space. The vehicles can be able to execute just a specific type of orders (dedicated vehicles [43]), as seen in [44], or the three types of orders (parallel vehicles [43]), as in the work of [45]. Usually, in the electric distribution maintenance/repair problem, the vehicles are characterized as parallel, i.e., the vehicle that attends a customer or maintenance order is the same that attends an emergency order, interrupting the predetermined route.

Given a set of orders (customer, maintenance, and emergency) to be serviced by a team, the dispatch center must define in real time a dispatch that covers all tasks in order to minimize the consumer's interruption time [36] and prioritize customers with higher loads (energy sales) and higher costs (tariffs and penalties).

Let us now define the difference between *offline* (or, *static*) and *online* (or, *dynamic*) dispatch. We will use here the definition by [6, 46–48].

7.4.2.1 Offline Dispatch

The dispatch is said *offline* when all problem inputs are received before dispatch determination and do not change thereafter.

The time efficiency of the *offline* algorithm is not a critical concern. In the *offline* dispatch, all time and resource constraints will have a better chance of being met.

Also, it is relatively easy to consider additional constraints, such as tasks arbitrary release times and vehicle capacity.

Most of the objective functions of *offline* problems, according to [46], are min (route cost, route distance, total lateness, number of vehicles, cost of service plus penalty, customer dissatisfaction, makespan) and max(quality of service, profit).

7.4.2.2 Online Dispatch

Research on online vehicle routing problems has grown considerably over the last decades or so. The survey of [49], catalogued some 154 references.

If new events occur over time in a dispatch, with the input on the problem received and updated concurrently with the determination of the dispatch (i.e., one or more problem parameters, like service and travel times, can be expressed as a function of time), the problem is characterized as *online*. In the *online* dispatch, no solution is constructed "a priori," because the decision-maker does not know in advance how many jobs are going to be processed and what their processing times are going to be. It is a policy that prescribes how the routes should evolve as a function of those inputs that evolve in real-time [50].

The *online* problem involves new elements that increase the complexity of their decisions thus, an *online* algorithm may fail to provide a feasible solution in the short computational time available. Additional constraints are likely to conflict with the priorities that are assigned at runtime to the processes. Compared to the *offline* dispatch, the dispatcher has more work to perform at runtime. On the other hand, *online* dispatch generally allows more flexible reactions in unforeseen situations than when operating *offline*. Due to the dynamic characteristics of the data, these problems require communication technologies (e.g., mobile phones, geographical information systems) that allow the exchange of data in real-time between the decision maker and the service teams.

According to [46], the objective function differs little between *online* and *offline* problems. However, some particularities can be explored in dealing with online problems. For the *stochastic online* problems, it is more frequent the use of "throughput" or "per unit time" objectives, such as average per unit time attended service order, average per unit time cost, average demand rejections per unit time, and so on. The service orders attendance that arises dynamically along the route course also introduces the notion of response time: a customer might request to be attended as soon as possible, in this case, the objective can be the latency minimization [49]. Also, the objective functions can be set to give more preference to events closer than to those that may occur later, due to other events that may occur in between [46] (see work of [51]).

In *online* dispatch, decisions are made in real-time as the scenario is updated. In this way, the quality of the decision (the more time available the better) competes with the speed of reactiveness to scenario changes (the faster the better).

The sources of dynamism can be various, but, according to [6, 36], the dynamic requests can be the service orders demands, being considered dynamically revealed, as in the case of emergency orders, with unforeseen character in space and time. The variability in travel times is also considered in studies, unlike the variable service times, addressed only by a few works. Other sources of dynamism, less predictable, are service cancellations, vehicle breakdowns, unexpected congestion and accidents, and unexpected changes in customer locations and demands.

Online dispatch problems can present different levels of dynamism according to their frequency of changes (the rate at which new information becomes available) and the urgency of requests (latency) [52]. Have a measure of the performance of the dispatch "dynamism" would be very valuable to examine the performance of an algorithm under varying conditions.

Larsen [53] proposed the *effective degree of dynamism* (*edod*) de, representing the normalized average of the disclosure times. In [54], the author extended the formulation to encompass problems with time window, representing the level of urgency of the requests.

Consider a planning horizon that starts at time 0 and finishes at time FT. Let T be the set of service orders and let t_i and l_i denote the time that service order $i \in T$ is known and the latest time request $i \in T$ can be served, respectively. The value t_i is sometimes called the *disclosure date* and it is equal to zero for any static request. The *edod* is defined in Eq. (7.32).

$$\text{edod} = \frac{1}{|T|} \sum_{i \in T} \frac{\text{FT} - (l_i - t_i)}{\text{FT}} \tag{7.32}$$

Larsen et al. [48, 54] use the *edod* to define a framework classifying *online* vehicle routing problem, defining as presented in Table 7.1. Observe that $0 \leq$ *edod* ≤ 1, being 0 for a purely static system and 1 for a purely dynamic system.

Based on the definitions of dynamism, Figure 7.4 provide a framework for classifying the service dispatch in the electric power systems according to its degree of dynamism and the system objective.

7.4.2.3 Dispatch Structure

Based on the above definition of *offline* and *online* scenario, the customer and maintenance orders can be characterized as an *offline* demand, while the emergency ones are characterized as an *online* demand. Figure 7.5 describes the timeline of and *offline* and an *online* dispatch, together with the events (static and dynamic orders, vehicle travel), the update and exchange of information between the vehicle and the dispatcher (GPS, GPRS, Dispatch center, status update times) and the type of

Table 7.1 Dynamic levels

edod	Dynamic level
<0.3	Weakly
0.3 $\leq$ 0.8	Moderately
0.8<	Strongly

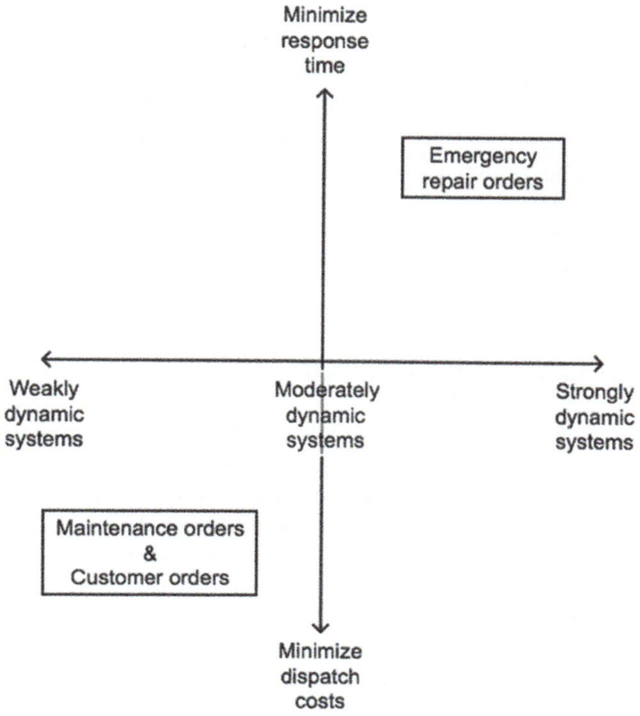

Fig. 7.4 Framework for classifying online dispatch problems in electric power systems by their degree of dynamism and their objective

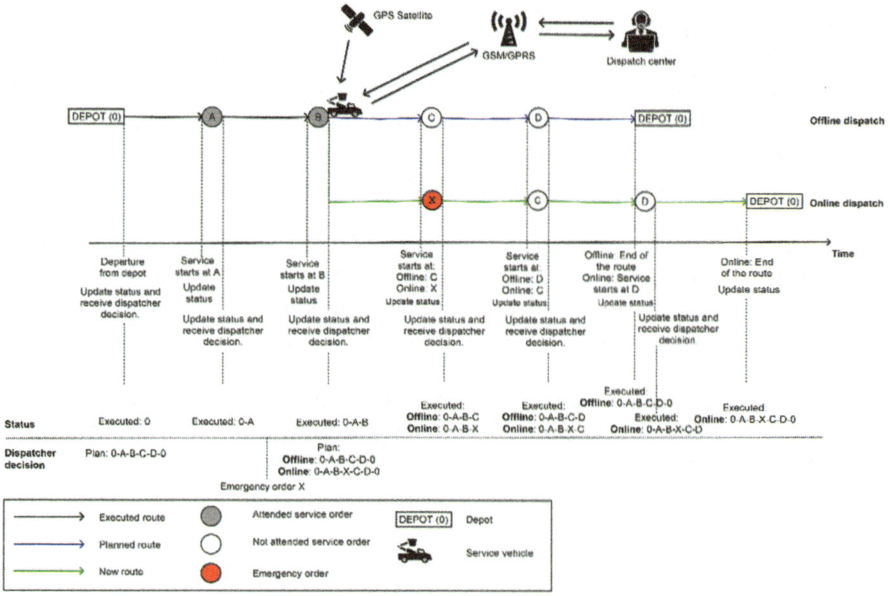

Fig. 7.5 Timeline of events and real-time communication for *offline* and *online* dispatches

information and decisions that are exchange between them (beginning of a service, end of service, executed orders, new route).

Following the definitions of [6, 46, 49], the *offline* service dispatch problem taxonomy can be set as *offline* and *deterministic*, or *offline* and *stochastic* dispatch problems. Also, the *online* (or, *dynamic*) service dispatch problem taxonomy can be set as *online* and *deterministic*, or *online* and *stochastic* dispatch problems. This taxonomy is defined in the following section.

7.4.2.4 Taxonomy of Dispatch Problems

In *offline* and *deterministic* dispatch problems, all the data (e.g., number of service orders, processing times, release dates, due dates, weights, number of vehicles, travel times, and so on) are assumed to be known in advance, and the route does not change once they are in execution. The decision-maker can then determine a priori the best service dispatch. This classical problem has been extensively studied in the literature (see [6, 55]).

In *offline* and *stochastic* dispatch problems, the data are affected by uncertainties, being partially known in the form of random variables, and the rest of data being revealed during the dispatch execution. It is assumed that the routes are defined a priori and only minor changes (travel time delay, skip a customer) are allowed afterwards. The most frequent uncertainties are when a customer needs to be serviced with a given probability, when either service or travel times have stochastic times, and when the model have stochastic demands. For more details see the works of [6, 56, 57].

In *online* and *deterministic* dispatch problems, it is assumed that part or all of the input is unknown a priori and revealed dynamically during the design or execution of the routes, becoming available over time (as addressed by [58–60]). No stochastic information (a probability) about the future, dynamically evolving inputs is known, the inputs values becomes known only when they appear [46] (e.g., nothing may be known about the location of a service order until the affected customers request service).

In *online* and *stochastic* dispatch problems, some probabilistic information about the inputs is known a priori (e.g., forecasts, probability distributions), updating the routes as these inputs evolve in time (e.g., the value of the service time, assumed to follow a certain probability distribution, is revealed when the vehicle arrive at the service order). In this set of problems, we try to explore, in the search for the solution, all the available predicted information about the future events, producing more efficient solutions than those obtained through myopic deterministic. This class of problems was addressed in the works of [6, 46, 47, 61–64].

7.4.3 Applications e Algorithms

Most of the algorithms used in dispatch are based on heuristics or metaheuristics, although exact methods are also utilized. Exact methods of a mathematical model operating together with heuristics or metaheuristics, called matheuristics, are also applied in the dispatch problem. We will now review some works on the dispatch in electric distribution systems.

Arab et al. [65] propose a decision-making tool to schedule the repair operations of transmission lines and buses in coordination with operations of its generation units in a cost-effective manner. The problem is formulated by a mixed-integer programming approach.

Nurre et al. [66] proposed an integrated network design and scheduling problem that models the restoration efforts. The author develops a heuristic of dispatching that selects the next set of tasks to be processed by the work groups. It is also proposed an integer programming formulation of which links the network design and scheduling decisions.

The work of Hentenryck [67] addresses in a decoupled way the problems of optimization of route re-planning and the restoration of the electric system through the method of constraint injection, in order to schedule and allocate the routes to fleets of repair crews, using the Neighborhood Search technique large-sized. This work was extended in Simon [68], where the joint repair/restoration problem has approached with a 3-stage decomposition, with the goal of minimizing the sum of the delivery times. A randomized adaptive vehicle decomposition was applied to scale to very large power networks restoration.

Garcia [45] proposes a mathematical model to schedule maintenance vehicles when considering emergency scenarios in electric distribution systems.

Perrier [36] presents a full survey of models and algorithms for emergency dispatch in electric distribution systems. In the work we can see that the strategies used are Composite heuristics; Depth-first search; Best-first search; Breadth-first search; Genetic algorithm; Neural network; Fuzzy logic; Simulated annealing; Ant colony optimization.

7.5 Conclusions

In this chapter we have studied the dispatch problem, its variants and application in the smart operation of power distribution systems. First, we have studied the anatomy of the dispatch problem and how it can be approached. Then, we have studied two instances of the dispatch problems faced by the distribution system operators: the economic dispatch of (distributed) power generators and the dispatch of crews for the service of maintenance and emergency orders. Indeed, we have observed that the research on the dispatch problems in distribution power systems has gained a lot of attention from the research community in the last years and believe that it will be of paramount importance for the operation of the future smart power distribution systems.

References

1. Dantzig GB, Ramser JH (1959) The truck dispatching problem. Manag Sci 6:80–91. https://doi.org/10.1287/mnsc.6.1.80
2. Lamorgese L, Mannino C (2013) The track formulation for the train dispatching problem. Electron Notes Discrete Math 41:559–566
3. Ruokokoski M, Sorsa J, Siikonen M-L, Ehtamo H (2016) Assignment formulation for the elevator dispatching problem with destination control and its performance analysis. Eur J Oper Res 252(2):397–406
4. Jagtenberg CJ, van den Berg PL, van der Mei RD (2017) Benchmarking online dispatch algorithms for emergency medical services. Eur J Oper Res 258(2):715–725
5. Burkard RE, Çela E (1999) Linear assignment problems and extensions. Handbook of combinatorial optimization, supplement, vol A. Kluwer, Dordrecht, pp 75–149. https://doi.org/10.1007/978-1-4757-3023-4_2
6. Toth P, Vigo D (2014) Vehicle routing: problems, methods, and applications. Vehicle routing. Second. [S.l.]: MOS-SIAM Series on Optimization.
7. Braekers K, Ramaekers K, Nieuwenhuyse IV (2016) The vehicle routing problem: state of the art classification and review. Comput Ind Eng 99:300–313. https://doi.org/10.1016/j.cie.2015.12.007
8. Lawler EL, Lenstra JK, Kan AHGR, Shmoys DB (1993) Chapter 9. Sequencing and scheduling: algorithms and complexity. In: Logistics of production and inventory. [S.l.]: Handbooks in operations research and management science, supplement C. Elsevier, New York, pp 445–522. https://doi.org/10.1016/S0927-0507(05)80189-6
9. Blazewicz J, Schmidt G, Shaw M (2007) Chapter 2. Basics. In: Logistics of production and inventory. [S.l.]. Handbook on scheduling: from theory to applications. Springer, Berlin/Heidelberg, pp 9–56
10. Blazewicz J, Schmidt G, Shaw M (2007) Chapter 3. Definition, analysis and classification of scheduling problems. In: Logistics of production and inventory. [S.l.]. Handbook on scheduling: from theory to applications. Springer, Berlin/Heidelberg, pp 9–56
11. Spliet R, Desaulniers G (2015) The discrete time window assignment vehicle routing problem. Eur J Oper Res 244(2):379–391. https://doi.org/10.1016/j.ejor.2015.01.020
12. Boonsam P, Suthikarnnarunai N, Chitphaiboon W (2011) Assignment problem and vehicle routing problem for an improvement of cash distribution. Proc World Congr Eng Comput Sci 2:1160–1164
13. Chen H-K, Hsueh C-F, Chang M-S (2009) Production scheduling and vehicle routing with time windows for perishable food products. Comput Oper Res 36(7):2311–2319. https://doi.org/10.1016/j.cor.2008.09.010
14. Ullrich CA (2013) Integrated machine scheduling and vehicle routing with time windows. Eur J Oper Res 227(1):152–165. https://doi.org/10.1016/j.ejor.2012.11.049
15. Desrosiers J, Dumas Y, Solomon MM, Soumis F (1995) Network routing. Handbook in operations research and management science. Chapter time constrained routing and scheduling, 8:35–139. North-Holland, doi:https://doi.org/10.1016/S0927-0507(05)80106-9
16. Chen Z-L, Pundoor G (2006) Order assignment and scheduling in a supply chain. Oper Res 54:555–572. https://doi.org/10.1287/opre.1060.0280
17. Mosheiov G, Oron D (2006) Due-date assignment and maintenance activity scheduling problem. Math Comput Model 44:1053–1057. https://doi.org/10.1016/j.mcm.2006.03.008
18. Hervert-Escobar L, López-Ramos F, Esquivel-Flores OA (2016) Integrated approach to assignment, scheduling and routing problems in a sales territory business plan. Procedia Comput Sci 80:1887–1896, International Conference on Computational Science 2016, ICCS 2016, 6–8, San Diego, CA. https://doi.org/10.1016/j.procs.2016.05.487
19. Happ HH (1977) Optimal power dispatch – a comprehensive survey. IEEE Trans Power Apparatus Syst PAS-96(3):841–854
20. Chowdhury BH, Rahman S A review of recent advances in economic dispatch. IEEE Trans Power Syst 5(4):1248–1259

21. Ipakchi A, Albuyeh F (2009) Grid of the future. Power Energy Mag 7:52–62
22. Cheung K, Wang X, Chiu B-C, Xiao Y, Rios-Zalapa R (2010) Generation dispatch in a smart grid environment. Innovative Smart Grid Technologies (ISGT), Gaithersburg, MD, pp 1–6
23. Basu M (2011) Economic environmental dispatch using multi-objective differential evolution. Appl Soft Comput 2(11):2845–2853
24. Varaiya PP, Wu FF, Bialek JW (2011) Smart operation of smart grid: risk-limiting dispatch. Proc IEEE 99(1):40–57. https://doi.org/10.1109/JPROC.2010.2080250
25. Li Q, Gao DW, Zhang H, Wu Z, Wang F-Y (2017) Consensus-based distributed economic dispatch control method in power systems. IEEE Trans Smart Grid PP(99):1. https://doi.org/10.1109/TSG.2017.2756041
26. Liang Y, Liu F, Mei S (2017) Distributed real-time economic dispatch in smart grids: a state-based potential game approach. IEEE Trans Smart Grid PP(99):1. https://doi.org/10.1109/TSG.2017.2652919
27. Ng SKK, Zhong J (2012) Smart dispatch of controllable loads with high penetration of renewables, PES T&D 2012, Orlando, FL, pp. 1–6, doi: https://doi.org/10.1109/TDC.2012.6281412
28. Xia X, Elaiw AM (2010) Optimal dynamic economic dispatch of generation: a review. Electr Power Syst Res 80(8):975–986. https://doi.org/10.1016/j.epsr.2009.12.012
29. Reddy SS, Bijwe PR, Abhyankar AR (2015) Real-time economic dispatch considering renewable power generation variability and uncertainty over scheduling period. IEEE Syst J 9 (4):1440–1451. https://doi.org/10.1109/JSYST.2014.2325967
30. Zhang Y, Rahbari-Asr N, Chow M-Y (2016) A robust distributed system incremental cost estimation algorithm for smart grid economic dispatch with communications information losses. J Netw Comput Appl 59:315–324. https://doi.org/10.1016/j.jnca.2015.05.014
31. Kar S, Hug G (2012) Distributed robust economic dispatch in power systems: A consensus + innovations approach. 2012 I.E. Power and Energy Society General Meeting, San Diego, CA, pp. 1–8. doi:https://doi.org/10.1109/PESGM.2012.6345156
32. Li N, Marden JR (2013) Designing games for distributed optimization. IEEE J Sel Top Signal Process 7(2):230–242. https://doi.org/10.1109/JSTSP.2013.2246511
33. Schäfer A, Moser A (2012) Dispatch optimization and economic evaluation of distributed generation in a virtual power plant, 2012 I.E. Energytech, Cleveland, OH, pp. 1–6. doi: https://doi.org/10.1109/EnergyTech.2012.6304655
34. Li P, Liu Y, Xin H, Jiang X (2017) A robust distributed economic dispatch strategy of virtual power plant under cyber-attacks. IEEE Trans Ind Inf PP(99):1. https://doi.org/10.1109/TII.2017.2788868
35. Graham RL, Lawler EL, Lenstra JK, Kan AHGR (1979) Optimization and approximation in deterministic sequencing and scheduling: a survey. In: Hammer PL, Johnson EL, Korte BH (eds) Discrete optimization II. [S.l.]: Annals of discrete mathematics, supplement C. Elsevier, New York, pp 287–326. https://doi.org/10.1016/S0167-5060(08)70356-X
36. Perrier N et al (2013) A survey of models and algorithms for emergency response logistics in electric distribution systems. Part II: contingency planning level. Comput Oper Res 40 (7):1907–1922. https://doi.org/10.1016/j.cor.2012.09.009
37. Guha S, Moss A, Naor JS, Schieber B (1999) Efficient recovery from power outage. In: Conference proceedings of the annual ACM symposium on theory of computing. Atlanta. New York, NY: GA: ACM; pp. 574–578. doi:https://doi.org/10.1145/301250.301406
38. Yao MJ, Min KJ (1998) Repair-unit location models for power failures. IEEE Trans Eng Manag 45:57–65. https://doi.org/10.1109/17.658661
39. Gillett BE, Miller LR (1974) A heuristic algorithm for the vehicle-dispatch problem. Oper Res 21:340–349. https://doi.org/10.1287/opre.22.2.340
40. Raff S (1983) Routing and scheduling of vehicles and crews: the state of the art. Comput Oper Res 10(2):63–212. https://doi.org/10.1016/0305-0548(83)90030-8
41. Miller CE, Tucker AW, Zemlin RA (1960) Integer programming formulations and travelling salesman problems. J Ass Comp Mach 7:326–329. https://doi.org/10.1145/321043.321046

42. Bektas T, Gouveia L (2014) Requiem for the Miller–Tucker–Zemlin subtour elimination constraints? Eur J Oper Res 236(3):820–832. https://doi.org/10.1016/j.ejor.2013.07.038
43. Blazewicz J, Schmidt G, Shaw M (2007) Chapter 7. Scheduling in hard real-time systems. In: Logistics of production and inventory. Handbook on scheduling: from theory to applications. Springer, Berlin/Heidelberg, pp 243–269
44. Zografos KG, Douligeris C, Tsoumpas P (1998) An integrated framework for managing emergency-response logistics: the case of the electric utility companies. IEEE Trans Eng Manag 45(2):115–126. https://doi.org/10.1109/17.669744
45. Garcia VJ et al (2014) Reliability assessment by coordinating maintenance vehicles in electric power distribution systems. Procedia Soc Behav Sci 111:1045–1053. https://doi.org/10.1016/j.sbspro.2014.01.139
46. Psaraftis HN, Wen M, Kontovas CA (2016) Dynamic vehicle routing problems: three decades and counting. Networks 67(1):3–31. https://doi.org/10.1002/net.21628
47. Ichoua S, Gendreau M, Potvin J-Y (2006) Exploiting knowledge about future demands for real-time vehicle dispatching. Transp Sci 40(2):211–225. https://doi.org/10.1287/trsc.1050.0114
48. Larsen A, Madsen OBG, Solomon MM (2007) Chapter 2. Classification of dynamic vehicle routing systems. In: Zeimpekis V, Tarantilis CD, Giaglis GM, Minis I (eds) Dynamic fleet management, Operations Research/Computer Science Interfaces Series, vol 38. Springer, New York, pp 19–40. https://doi.org/10.1007/978-0-387-71722-7_2
49. Pillac V, Gendreau M, Guéret C, Medaglia AL (2013) A review of dynamic vehicle routing problems. Eur J Oper Res 225:1–11. https://doi.org/10.1016/j.ejor.2012.08.015
50. Psaraftis HN (1988) Dynamic vehicle routing. In: Golden BL, Assad AA (eds) Vehicle routing: methods and studies. North-Holland, Amsterdam, pp 223–248. https://doi.org/10.1002/net.21628
51. Bertsekas DP (2012) Dynamic programming and optimal control: approximate dynamic programming, vol II, 4th edn. Athena Scientific, Belmont, MA
52. Ichoua S, Gendreau M, Potvin J-Y (2007) Chapter 1. Planned route approaches for real-time vehicle routing. In: Zeimpekis V, Tarantilis CD, Giaglis GM, Minis I (eds) Dynamic fleet management: concepts, systems, algorithms and case studies, Operations Research/Computer Science Interfaces Series, vol 38. Springer, New York, pp 1–18. https://doi.org/10.1007/978-0-387-71722-7_1
53. Larsen A (2001) The dynamic vehicle routing problem, PhD thesis, Technical University of Denmark (DTU)
54. Larsen A, Madsen OBG, Solomon MM (2002) Partially dynamic vehicle routing models and algorithms. J Oper Res Soc 53(6):637–646. https://doi.org/10.1057/palgrave.jors.2601352
55. Laporte G (2009) Fifty years of vehicle routing. Transp Sci 43(4):408–416. https://doi.org/10.1287/trsc.1090.0301
56. Bertsimas D, Simchi-Levi D (1996) A new generation of vehicle routing research: robust algorithms, addressing uncertainty. Oper Res 44:286–304. https://doi.org/10.1287/opre.44.2.286
57. Gendreau M, Laporte G, Séguin R (1996) Stochastic vehicle routing. Eur J Oper Res 88:3–12. https://doi.org/10.1016/0377-2217(95)00050-X
58. Psaraftis HN (1980) A dynamic-programming solution to the single vehicle many-to-many immediate request dial-a-ride problem. Transp Sci 14:130–154. https://doi.org/10.1287/trsc.14.2.130
59. Ichoua S, Gendreau M, Potvin J-Y (2003) Vehicle dispatching with time-dependent travel times. Eur J Oper Res 144(2):379–396. https://doi.org/10.1016/S0377-2217(02)00147-9
60. Gendreau M, Guertin F, Potvin J-Y, Séguin E (2006) Neighborhood search heuristics for a dynamic vehicle dispatching problem with pick-ups and deliveries. Transport Res Part C: Emerg Technol 14:157–174. https://doi.org/10.1016/j.trc.2006.03.002
61. Ferrucci F, Block S, Gendreau M (2013) A pro-active real- time control approach for dynamic vehicle routing problems dealing with the delivery of urgent goods. Eur J Oper Res 225:130–141. https://doi.org/10.1016/j.ejor.2012.09.016

62. Goodson JC, Jeffrey WO, Thomas BW (2013) Rollout policies for dynamic solutions to the multivehicle routing problem with stochastic demand and duration limits. Oper Res 61:138–154. https://doi.org/10.1287/opre.1120.1127
63. Thomas BW, White CCI (2004) Anticipatory route selection. Transp Sci 38:473–487. https://doi.org/10.1287/trsc.1030.0071
64. Jaillet P, Wagner MR (2008) Generalized online routing: new competitive ratios, resource augmentation, and asymptotic analyses. Oper Res 56(3):745–757. https://doi.org/10.1287/opre.1070.0450
65. Arab A, Khodaei A, Khator SK, Han Z (2016) Electric power grid restoration considering disaster economics. IEEE Access 4:639–649. https://doi.org/10.1109/ACCESS.2016.2523545
66. Nurre SG et al (2012) Restoring infrastructure systems: an integrated network design and scheduling (INDS) problem. Eur J Oper Res 223(3):794–806. https://doi.org/10.1016/j.ejor.2012.07.010
67. Hentenryck PV, Coffrin C, Bent R (2011) Vehicle Routing for the Last Mile of Power System Restoration. 17th Power Systems Computation Conference, v. 836
68. Simon B, Coffrin C, Hentenryck PV (2012) Chapter 25. Randomized adaptive vehicle decomposition for large-scale power restoration, computer science. In: Integration of AI and OR techniques in contraint programming for combinatorial optimzation problems. Springer, Berlin/Heidelberg, pp 379–394. https://doi.org/10.1007/978-3-642-29828-8_25

Index

D. P. Bernardon, V. J. Garcia (eds.), *Smart Operation for Power Distribution Systems*, https://doi.org/10.1007/978-3-319-93922-3

R

S